高职高专国家骨干院校
重点建设专业(机械类)核心课程"十二五"规划教材

单片机实训教程

主 编 李庭贵 龚勤慧 张化锦

合肥工业大学出版社

图书在版编目(CIP)数据

单片机实训教程/李庭贵,龚勤慧,张化锦主编. —合肥:合肥工业大学出版社,2012.8
ISBN 978-7-5650-0733-0

Ⅰ.①单… Ⅱ.①李… Ⅲ.①单片机微型计算机——高等学校—教材 Ⅳ.①TP368.1

中国版本图书馆 CIP 数据核字(2012)第 102412 号

单片机实训教程

李庭贵 龚勤慧 张化锦 主编		责任编辑 马成勋
出 版	合肥工业大学出版社	版 次 2012 年 8 月第 1 版
地 址	合肥市屯溪路 193 号	印 次 2012 年 8 月第 1 次印刷
邮 编	230009	开 本 787 毫米×1092 毫米 1/16
电 话	总 编 室:0551—2903038	印 张 18
	市场营销部:0551—2903198	字 数 438 千字
网 址	press.hfut.edu.cn www.hfutpress.com.cn	印 刷 中国科学技术大学印刷厂
E-mail	hfutpress@163.com	发 行 全国新华书店

ISBN 978-7-5650-0733-0　　　　　定价:38.00 元
如果有影响阅读的印装质量问题,请与出版社发行部联系调换。

前　言

单片机应用技术是自动化、机电、应用电子技术等专业的核心课程，是为从事电子产品生产与开发企业培养具有单片机应用产品设计、分析、调试与制作能力的实践人才，它对学生职业岗位能力培养和职业素质培养起主要支撑作用。

单片机的特点及优势决定了其应用的广泛性，其智能化的潜力、较强接口驱动能力及低功耗的特点，广泛应用于工业领域、信息领域和家电领域，而且其技术日新月异。本教材的编写思路是便于学生入门，不仅仅是一本实训教程，同时也是一本单片机应用技术教材。

高职教育强调"以能力为本位，以职业实践为主线，努力做到把理论知识嵌入实践教学中"。该教材充分体现了任务引领与实践导向课程设计的思想，以理论与实践相结合为主线，使学生学习之后能够轻松地掌握单片机的基础知识，并具有初步开发、设计单片机产品的能力。把教学过程的"教、学、做"有效地融为一体，具有鲜明的高职教材特色。

本教材按照任务驱动的课程体系编写，总共包含25个任务，任务1 WAVE软件的使用，任务2 KEIL C软件的使用，任务3 EASY 51PRO 烧写软件，任务4 C51程序设计知识，任务5 闪烁灯控制，任务6 模拟开关灯控制，任务7 多路开关状态指示灯控制，任务8 广告灯控制，任务9 流水灯控制，任务10 数码管显示技术，任务11 动态数码显示技术，任务12 独立按键识别技术，任务13 一键多功能按键识别技术，任务14 矩阵键盘识别技术，任务15 字符型LCD显示，任务16 按键变量加减LCD显示，任务17 报警产生器，任务18 外部计数器中断，任务19 音乐发生器设计，任务20 马表设计，任务21 RS—232 串行通信，任务22 A/D转换，任务23 D/A转换，任务24 I^2C 总线存储器读写，任务25 DS18B20温度控制，同时给出各任务相应的电路原理图和参考程序（包括汇编源程序和C语言源程序）。每个任务包括实训任务、实训设备、硬件设计、软件设计、实训考核等部分，在任务驱动教学中包含了"单片机实训教学大纲"规定应掌握的所有知识点。

本书可作为高职高专自动化、机电、应用电子技术等专业的单片机实训教材，也可作为中等专业相关专业及职业培训的教材，还可以做为单片机技术爱好者参考用书。

本书由泸州职业技术学院李庭贵、龚勤慧、张化锦共同主编完成，其中龚勤慧编写了任务3、任务5、任务8、任务10、任务11、任务13，其余由李庭贵编写，全书由李庭贵统稿。成都庆丰电子工作室的李国庆工程师为本书的编写提供了大力的帮助，在此表示衷心感谢。

由于编者水平有限，难免会有错误与不妥之处，恳请广大读者批评指正。

编　者

2012年08月

目　录

任务 1　Wave 软件的使用 ……………………………………………………………… (1)

任务 2　Keil C 软件的使用 ……………………………………………………………… (8)

任务 3　Easy 51Pro 烧写软件 …………………………………………………………… (15)

任务 4　C51 程序设计知识 ……………………………………………………………… (20)

任务 5　闪烁灯控制 ……………………………………………………………………… (37)

任务 6　模拟开关灯控制 ………………………………………………………………… (44)

任务 7　多路开关状态指示灯控制 ……………………………………………………… (49)

任务 8　广告灯控制 ……………………………………………………………………… (55)

任务 9　流水灯控制 ……………………………………………………………………… (61)

任务 10　数码管显示技术 ……………………………………………………………… (67)

任务 11　动态数码显示技术 …………………………………………………………… (74)

任务 12　独立按键识别技术 …………………………………………………………… (80)

任务 13　一键多功能按键识别技术 …………………………………………………… (86)

任务 14　4×4 矩阵键盘识别技术 ……………………………………………………… (93)

任务 15　字符型 LCD 显示 ……………………………………………………………… (102)

任务 16　按键变量加减 LCD 显示 ……………………………………………………… (118)

任务 17　报警产生器 …………………………………………………………………… (128)

任务 18　外部计数器中断 ……………………………………………………………… (136)

任务 19　音乐发生器设计 ……………………………………………………………… (148)

任务 20　99.9 秒马表设计 ……………………………………………………………… (159)

任务 21　RS－232 串行通信 ·· (170)

任务 22　A/D 转换 ··· (186)

任务 23　D/A 转换 ··· (203)

任务 24　I^{2}C 总线存储器读写 ··· (224)

任务 25　DS18B20 温度控制 ·· (252)

附录 A　特殊功能寄存器 ·· (270)

附录 B　MCS－51 单片机指令系统 ··· (271)

附录 C　单片机伪指令 ··· (279)

参考文献 ··· (281)

任务1 Wave软件的使用

一、仿真器连接

仿真器的连接如图1-1所示。

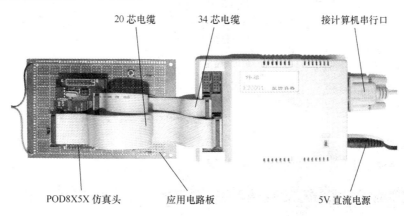

图1-1 仿真器的连接

二、仿真器设置

仿真器的设置菜单如图1-2所示。

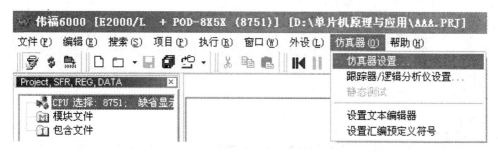

图1-2 仿真器的设置菜单

三、程序编译调试

1. 新建文件

选择菜单"文件"→"新建文件"如图1-3所示。

伟福6000：软件模拟器
文件(F) 编辑(E) 搜索(S)
打开文件(O) F3
保存文件(S) F2
新建文件(N)
另存为(A)...
重新打开
打开项目...
保存项目
新建项目...
关闭项目
项目另存为...
复制项目...
调入目标文件...
保存目标文件...
反汇编...
退出(X)

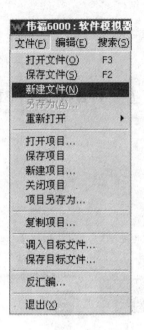

图 1-3 新建文件

2. 输入程序

在源程序窗口中输入所编写的程序，如图 1-4 所示。

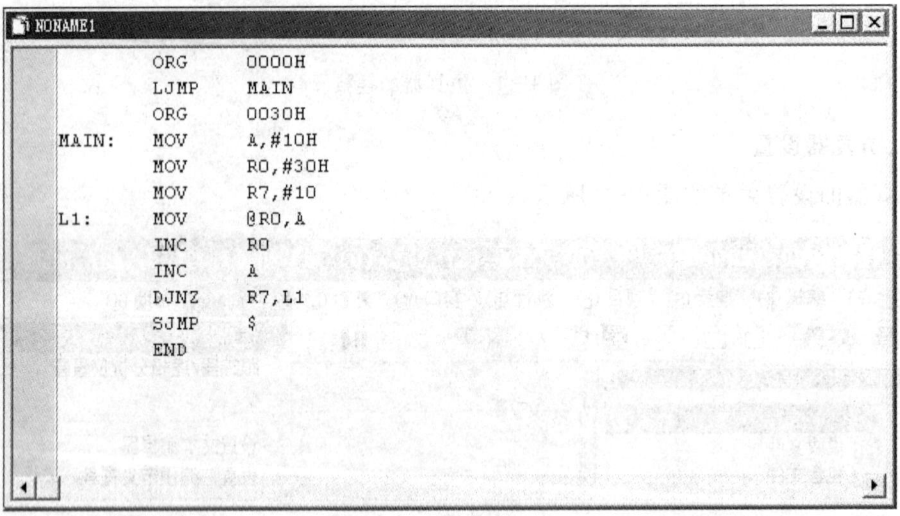

NONAME1
ORG 0000H
LJMP MAIN
ORG 0030H
MAIN: MOV A,#10H
MOV R0,#30H
MOV R7,#10
L1: MOV @R0,A
INC R0
INC A
DJNZ R7,L1
SJMP $
END

图 1-4 输入程序

3. 保存程序

选择菜单“文件”→“保存文件”，保存时文件名称必须带上后缀名“.ASM”，如图 1-5 所示。

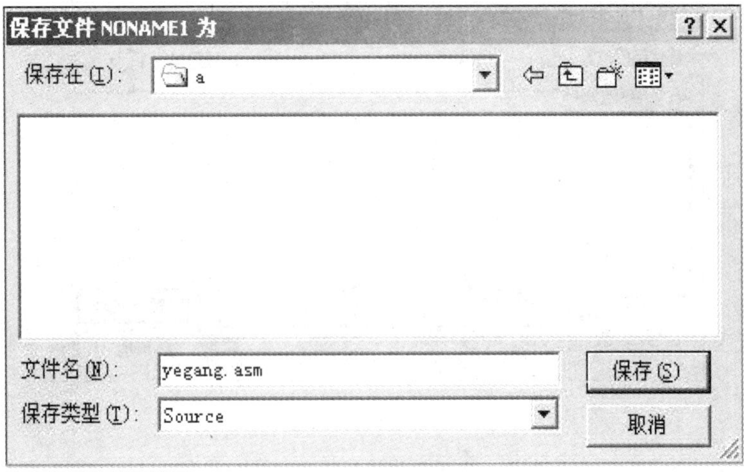

任务1 Wave软件的使用

图1-5 保存程序

4．建立新项目

选择菜单"文件"→"新建项目"，如图1-6所示。

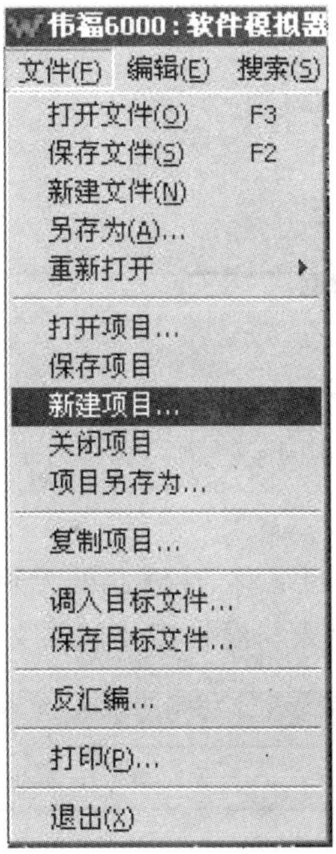

图1-6 建立新项目

5．加入模块文件

在弹出的窗口中，加入模块文件，选择刚才保存的文件 YEGANG.ASM，如图1-7所示。

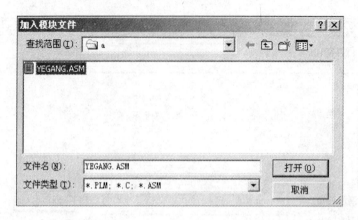

图1-7 加入模块文件

6. 加入包含文件

加入包含文件,若没有包含文件,则可按取消键。此处按取消键,如图1-8所示。

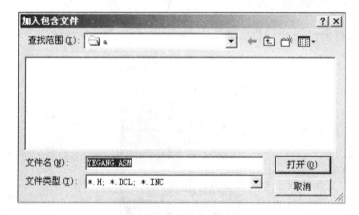

图1-8 加入包含文件

7. 保存项目

在"保存项目"对话框中输入项目名称,注意此处无须添加后缀名,软件会自动将后缀名设成".PRJ"。按"保存"键将项目存入与源程序相同的文件夹下,如图1-9所示。

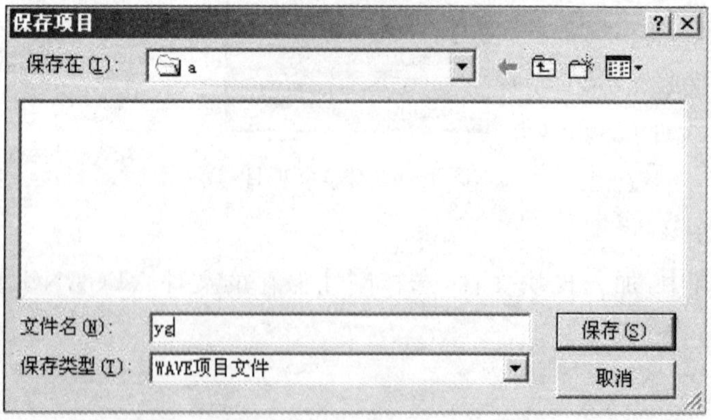

图1-9 保存项目

8. 仿真器设置

选择菜单"设置"→"仿真器设置",在弹出的"仿真器设置"对话框中,按如图 1-10 所示的设置,选择"使用伟福软件模拟器"。

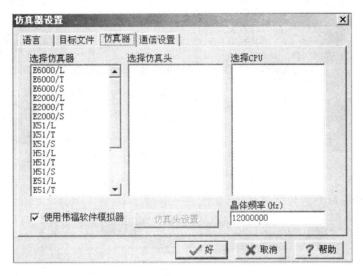

图 1-10 仿真器设置

9. 程序编译

选择菜单"项目"→"编译",进行程序编译,如图 1-11 所示。

如程序正确,编译后将产生两种格式的目标文件:二进制格式(BIN)目标文件和英特尔格式(HEX)目标文件,如图 1-11 所示。如果程序有错误,则将在信息窗口指出错误指令所在的源程序、行号、错误代码及错误原因,应先修改程序,然后再重新进行程序编译。

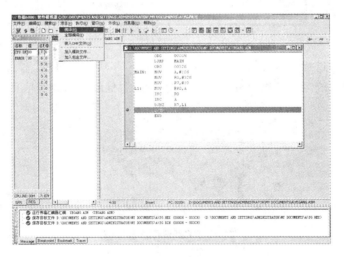

图 1-11 程序编译

10. 程序调试

双击"项目"中的 YEGANG.ASM 文件,在所需设置断点的指令上用右键设置断点,如图 1-12 所示。

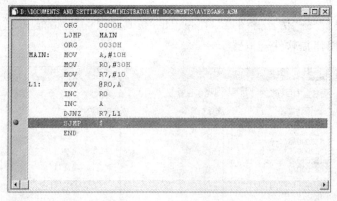

图 1-12 程序调试

在程序调试时,可选择菜单"执行"→"全速执行"、"执行|跟踪"、"执行|单步"、"执行"→"执行到光标处"等命令运行调试程序,如图 1-13 所示。

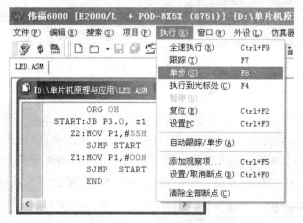

图 1-13 "执行"菜单

在调试运行的过程中,打开"窗口"菜单,如图 1-14 所示,可以选择打开 CPU 窗口、数据窗口等以观察系统运行过程中有关寄存器的状态、存储器的内容等信息,从而判断程序执行是否正确。

图 1-14 "窗口"菜单

11. 查看结果

选择菜单[窗口|数据窗口|DATA]，如图 1-15 所示。

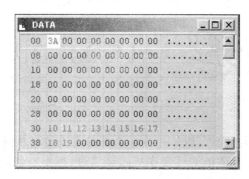

图 1-15　查看结果

注意：

DATA——片内 RAM 区域。

CODE——ROM 区域。

XDATA——片外 RAM 区域。

PDATA——分页式数据存储器，51 系列单片机中不用。

BIT——位寻址区域。

任务2 Keil C 软件的使用

随着单片机技术的不断发展,以单片机 C 语言为主流的高级语言也不断被更多的单片机爱好者和工程师所喜爱。使用 C51 要用到编译器,以便把写好的 C 程序编译为机器码,这样单片机才能执行编写好的程序。

KEIL μVision2 是众多单片机应用开发软件中优秀的软件之一,它支持众多公司的 MCS51 架构的芯片,它集编辑、编译、仿真等于一体,同时还支持 PLM、汇编和 C 语言的程序设计,它的界面和常用的微软 VC++ 的界面相似,界面友好,易学易用,在调试程序、软件仿真方面也有很强大的功能。

下面介绍 Keil C51 软件的使用方法。

进入 Keil C51 后,启动界面如图 2-1 所示。

图 2-1 Keil C51 的启动界面

几秒钟后出现编辑界面,如图 2-2 所示。

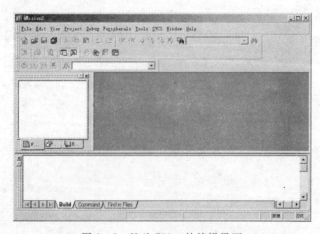

图 2-2 Keil C51 的编辑界面

学习程序设计语言及某种程序软件,最好的方法是直接操作实践。下面通过简单的编程、调试,引导大家学习 Keil C51 软件的基本使用方法和基本的调试技巧。

(1)建立一个新工程,单击"Project"菜单,在弹出的下拉菜单中选中"New Project"选项,如图 2-3 所示。

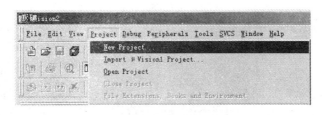

图 2-3 【New Project】菜单

(2)然后选择你要保存的路径,输入工程文件的名字,比如保存到 C51 目录里,工程文件的名字为 C51,如图 2-4 所示,然后点击保存。

图 2-4 保存文件窗口

(3)这时会弹出一个对话框,要求选择单片机的型号,可以根据你使用的单片机来选择,keil c51 几乎支持所有的 51 核的单片机,本书以应用广泛的 Atmel 的 89C51 为例说明,如图 2-5 所示。选择 89C51 之后,右边栏是对这个单片机的基本的说明,然后点击确定。

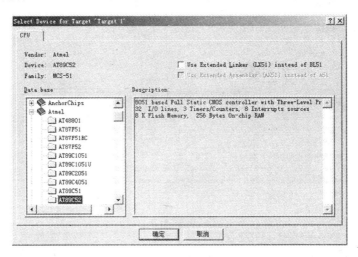

图 2-5 选择芯片型号

(4)完成上一步骤后,屏幕如图2-6所示。

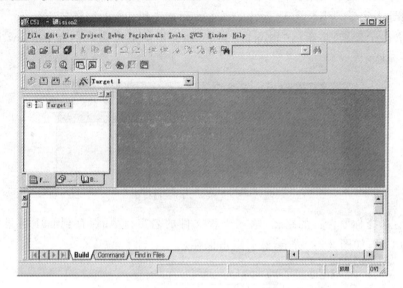

图2-6　Keil C51的编程界面

下面开始编写第一个程序。

(5)如图2-7所示,单击"File"菜单,再在下拉菜单中单击"New"选项。

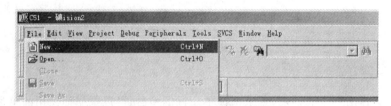

图2-7　"File"菜单的"New"选项

新建文件后屏幕如图2-8所示。

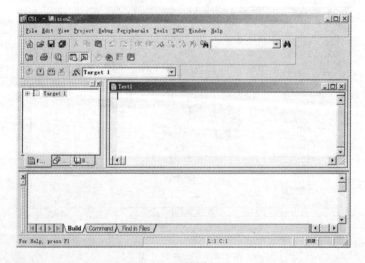

图2-8　编辑窗口

此时光标在编辑窗口里闪烁,这时可以键入用户的应用程序了,但应首先保存该空白的文件,单击菜单上的"File",在下拉菜单中选中"Save As"选项单击,如图 2-9 所示,在"文件名"栏右侧的编辑框中,键入欲使用的文件名,同时,必须键入正确的扩展名。注意,如果用 C 语言编写程序,则扩展名为(.c);如果用汇编语言编写程序,则扩展名必须为(.asm)。然后,单击"保存"按钮。

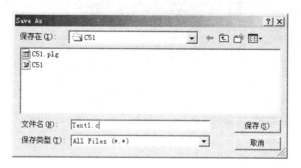

图 2-9 "Save As"选项设置

(6)回到编辑界面后,单击"Target 1"前面的"+"号,然后在"Source Group 1"上单击右键,弹出菜单如图 2-10 所示:

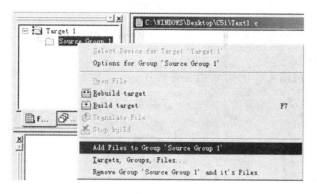

图 2-10 "添加文件"选择

然后单击"Add File to Group 'Source Group 1'"屏幕如图 2-11 所示。

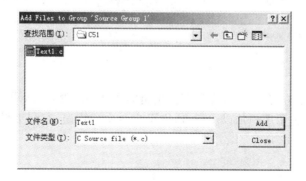

图 2-11 "添加文件"窗口

选中 Test.c,然后单击"Add",屏幕如图 2-12 所示。

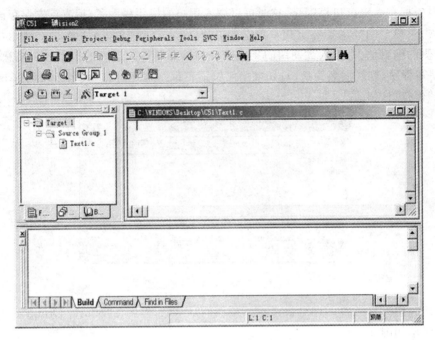

图 2-12 "添加文件"成功窗口

注意到"Source Group1"文件夹中多了一个子项"Text1.c"了吗？子项的多少与所增加的源程序的多少相同。

(7)编辑程序。请输入如下的 C 语言源程序：

```
#include<reg52.h>              //包含单片机头文件
#include<stdio.h>
void main(void)                //主函数
{
    SCON = 0x52;
    TMOD = 0x20;
    TH1 = 0xf3;
    TR1 = 1;                   //此行及以上 3 行为 printf 函数所必须
    printf("Hello I am KEIL. \n");  //打印程序执行的信息
    printf("I will be your friend. \n");
    while(1);
}
```

在输入上述程序时，会理解事先保存待编辑的文件的好处，即 Keil c51 会自动识别关键字，并以不同的颜色提示用户加以注意，这样用户会少犯错误，有利于提高编程效率。程序输入完毕后，如图 2-13 所示。

任务2　Keil C软件的使用

图 2-13　程序编辑界面

(8)编译程序。在图 2-13 中,单击"Project"菜单,再在下拉菜单中单击"Built Target"选项(或者使用快捷键 F7),编译成功后,再单击"Project"菜单,在下拉菜单中单击"Start/Stop Debug Session"(或者使用快捷键 Ctrl+F5),屏幕如图 2-14 所示。

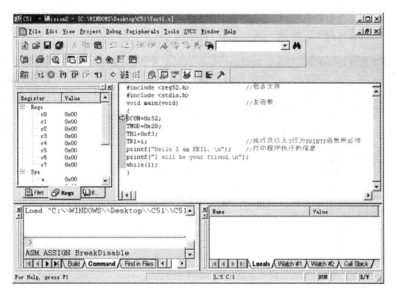

图 2-14　编译程序

(9)调试程序。在图 2-14 中,单击"Debug"菜单,在下拉菜单中单击"Go"选项,(或者使用快捷键 F5),然后再单击"Debug"菜单,在下拉菜单中单击"Stop Running"选项(或者使用快捷键 Esc);再单击"View"菜单,再在下拉菜单中单击"Serial Windows #1"选项,就可以看到程序运行后的结果,其结果如图 2-15 所示。

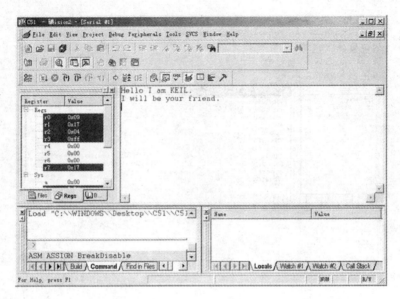

图 2-15 调试程序

至此,我们在 Keil C51 上做了一个完整工程项目。但这只是纯软件的开发过程,如何使用程序下载器看一看程序运行的结果呢?

(10)单击"Project"菜单,再在下拉菜单中单击"Options for Target 'Target 1'",在图 2-16 中,单击"Output"中单击"Create HEX File"选项,使程序编译后产生 HEX 代码,供下载器软件使用。把程序下载到 AT89S51 单片机中。

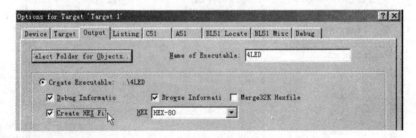

图 2-16 下载设置

任务 3 Easy 51Pro 烧写软件

一、软件使用

AT89S52 单片机烧写软件 Easy 51Pro v2.0 是专门用于下载程序到单片机系统中,该软件使用方便。启动软件之后进入以下界面,如图 3-1 所示。

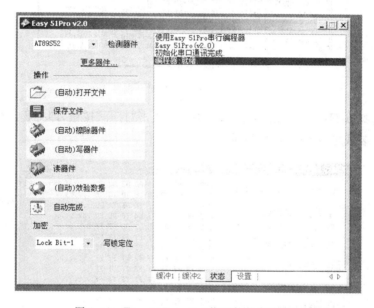

图 3-1 Easy 51Pro v2.0 烧写软件启动界面

系统显示"就绪"就说明已和实验板连接成功了,否则请重新检查串口连接线和电源连线。

(1)擦除:是把单片机的内容擦除干净,即单片机内部 ROM 的内容全为 FFH。

(2)写器件:把缓冲 1 区中的程序代码下载到单片机的内部 ROM 中。注意在编程之前,要对单片机芯片进行擦除操作。

(3)读器件:从单片机内部 ROM 中读取内容到代码缓冲 2 区中。

(4)效验数据:是经过编程之后,对下载到单片机内部 ROM 中的内容与代码区的内容相比较,若程序下载过程中完全正确,则提示校验正确,否则提示出现错误,需要重新下载程序到 ROM 中。

(5)自动:提供了从内部 ROM 从擦除到编程,最后到校验这三个过程。

(6)打开文件:把经过 KEIL C 软件转化成 HEX 格式的文件装入缓冲 1 中,当单击"装载"按钮时,出现如图 3-2 所示的对话框。

图 3-2　打开文件对话框

在这里选择以 HEX 为后缀的文件,选中它并点击"打开"按钮,即把程序代码装入到代码缓冲 1 中。装载之后如图 3-3 所示。

图 3-3　编程器缓冲区

我们就可以把代码缓冲 1 中的代码通过 ISP 方式下载到 AT89S52 单片机中。

(7)设置:对该软件一些操作方式进行设置,点击按钮之后,出现如图 3-4 所示的界面。

任务3　Easy 51Pro 烧写软件

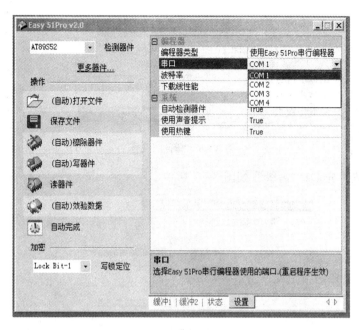

图 3-4　编程器设置

在这里可以进行通信端口的设置，共设置的 4 个串行通信端口，COM1、COM2、COM3、COM4，根据计算机的硬件特点来决定，默认情况下为 COM1，即串行通信口 1。

在进行程序调试的时候，一般通过 KEIL C 软件把编译好的程序转化成 HEX 格式文件，通过上面的方法，装载程序之后，点击"自动"按钮，程序就下载到单片机内部 ROM 芯片中，即可以看到程序的结果。

二、USB 转串口线安装和使用

如你的电脑没有串行接口，那就需要配购一条 USB 转串口线来下载程序。下面介绍一下它的安装步骤和使用方法。

第一步，驱动程序安装：PL－2303 Driver Install。

第二步，查找虚拟串口：在"控制面板"→"系统"→"设备管理"，可以看到现在 USB→232 是转到哪一个串口的。

如图 3-5 所示，虚拟串口就是 COM4。如果现在虚拟串口号大于 COM4，点击鼠标右键，选择"属性"，如图 3-6 所示。

图 3-5　查找虚拟串口

图 3-6 "属性"选择

在端口设置里先选择"高级",如图 3-7 所示。

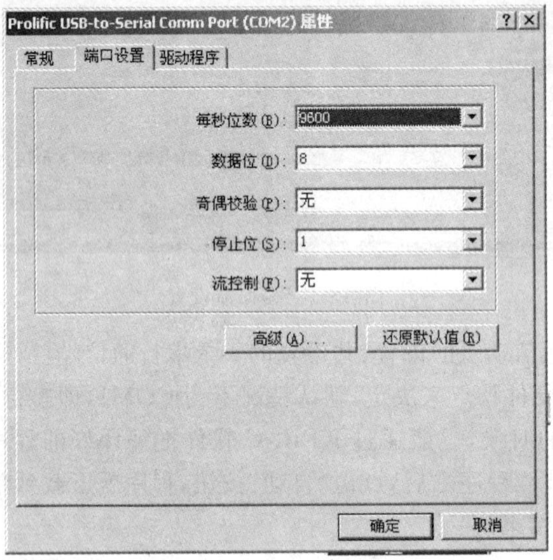

图 3-7 "高级"设置

在端口号处选 COM2~COM4,强制指定串口,因为 EASY 51PRO 软件只支持 COM1~COM4,如图 3-8 所示。

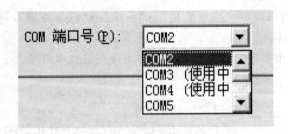

图 3-8 强制指定串口

第三步,运行 EASY 51PRO 烧写软件,在"设置"选项里把串口改成相对应虚拟串口,如图 3-9 所示,然后关闭软件。

任务3 Easy 51Pro 烧写软件

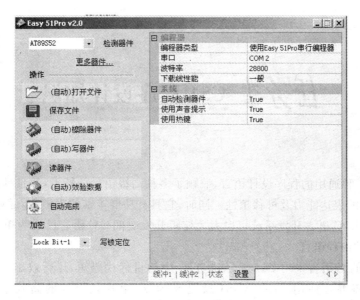

图 3-9 编程器设置信息显示

第四步，再重新启动 EASY 51PRO 烧写软件，显示"就绪"就可以了，如图 3-10 所示。

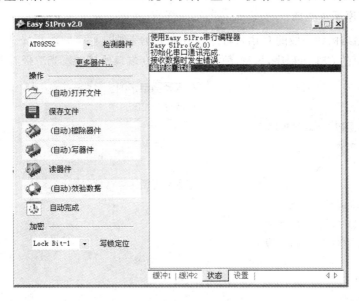

图 3-10 编程器就绪

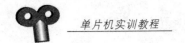

任务 4 C51 程序设计知识

C 语言是一种通用的程序设计语言,兼顾了多种高级语言的特点,并且具备汇编语言的功能,具有很强的表达能力及可移植性。同时,它具有功能丰富的库函数,运算速度快、编译效率高。因此,用 C 语言开发系统可以大大缩短开发时间。C 语言已成为单片机开发过程中应用最广泛的编程语言。

51 单片机的 C 语言采用 C51 编译器,它产生的目标代码短、运行效率高,可与 A51 汇编语言目标代码混合使用。Keil Cx51 就是专为 51 单片机设计的高效率编译器,符合 ANSI 标准,所需的存储空间极小。编写好的 C 语言源程序经过 C51 编译器编译、L51 连接定位后就可以生产目标程序,通过编程器可以下载到 51 单片机中。

一、C51 程序基本构成

C51 源程序结构和一般的 C 语言程序没有太大的差别。C51 的源程序文件扩展名为 .c。下面是一个简单的 C51 源程序,该程序可以实现 P0.0 端口所接的发光二极管闪烁点亮。

```
#include<AT89X51.H>          //编译器自带的头文件
sbit L1=P0^0;                //定义 P0.0
//延时 0.2s 函数 delay02s()
void delay02s(void)
{
    unsigned char i,j,k;     //定义无符号字符型变量 i、j、k
    for(i=20;i>0;i--)
        for(j=20;j>0;j--)
            for(k=230;k>0;k--);
}
//主函数
void main(void)
{
    while(1)
    {
        L1=0;                //当 P0.0=0 时,发光二极管 L1 点亮
        delay02s();          //调用函数 delay02s()
        L1=1;                //当 P0.0=1 时,发光二极管 L1 熄灭
        delay02s();          //调用函数 delay02s()
    }
}
```

一个C51源程序是一个函数的集合,函数是C程序的基本单位。在这个集合中,至少包括一个主函数main( ),也可以包含一个主函数和若干其他的函数,其他函数都可以被主函数调用,也可以相互调用,但main( )函数不能被其他函数调用。被调用的函数可以是编译器提供的库函数,也可以是用户自己编制设计的函数。

不论主函数main( )在什么位置,程序的执行都是从主函数main( )开始执行的,主函数是程序的入口,主函数中的语句执行完毕,则程序执行结束。

C51源程序书写格式自由,一行可以书写多条语句,一个语句也可以分多行书写。但在每个语句和数据定义的最后必须有一个分号。分号是C语句的必要组成部分,分号必不可少,即使程序中最后一个语句也应该包含分号。

C51程序可以用/*……*/或//对C51源程序中的任何部分作注释。一个好的有价值的程序都应该加上必要的注释,以增加可读性。

二、C51数据结构

数据:具有一定格式的数字或数值叫做数据。数据是计算机操作的对象,不管使用任何语言何种算法进行程序设计,在计算机中运行的只有数据流。

数据类型:数据不同的格式叫数据类型。

数据结构:数据按一定的数据类型进行排列、组合、架构称为数据结构。

C51的数据结构如表4-1所示。

表4-1 C51的数据结构

数据类型	长 度	值 域
unsigned char	单字节8位	0~255
signed char	单字节8位	-128~+127
unsigned int	双字节16位	0~65535
signed int	双字节16位	-32768~+32767
unsigned long	四字节32位	0~4294967295
signed long	四字节32位	-2147483648~+2147483647
float	四字节32位	±1.175494E-38±3.402823E+38(6位数字)
double	八字节64位	±1.175494E-38~±3.402823E+38(10位数字)
*	1~3字节	对象的地址
bit	位	0或1
sfr	单字节	0~255
sfr16	双字节	0~65535
sbit	位	0或1

1. Char:字符类型

char类型的长度是一个字节,通常用于定义处理字符数据的变量或常量。分无符号字符类型unsigned char和有符号字符类型signed char,默认值为signed char类型。

unsigned char 类型用字节中所有的位来表示数值,所可以表达的数值范围是 0~255。unsigned char 常用于处理 ASCII 字符或用于处理小于或等于 255 的整型数。

signed char 类型用字节中最高位字节表示数据的符号,"0"表示正数,"1"表示负数,负数用补码表示。所能表示的数值范围是-128~+127。

正数的补码与原码相同,负二进制数的补码等于它的绝对值按位取反后加 1。

2. int:整型

int 整型长度为两个字节,用于存放一个双字节数据。分有符号 int 整型数 signed int 和无符号整型数 unsigned int,默认值为 signed int 类型。

signed int 表示的数值范围是-32768~+32767,字节中最高位表示数据的符号,"0"表示正数,"1"表示负数。

unsigned int 表示的数值范围是 0~65535。

3. long:长整型

long 长整型长度为四个字节,用于存放一个四字节数据。分有符号长整型 signed long 和无符号长整型 unsigned long,默认值为 signed long 类型。

signed int 表示的数值范围是-2147483648~+2147483647,字节中最高位表示数据的符号,"0"表示正数,"1"表示负数。

unsigned long 表示的数值范围是 0~4294967295。

4. float:浮点型

float 浮点型在十进制中具有 7 位有效数字,是符合 IEEE-754 标准的单精度浮点型数据,占用四个字节。

5. *:指针型

指针型本身就是一个变量,在这个变量中存放的指向另一个数据的地址。这个指针变量要占据一定的内存单元,对不同的处理器长度也不尽相同,在 C51 中它的长度一般为 1~3 个字节。指针变量也具有类型。

6. bit:位类型

bit 位标量是 C51 编译器的一种扩充数据类型,利用它可定义一个位标量,但不能定义位指针,也不能定义位数组。它的值是一个二进制位,不是 0 就是 1,类似一些高级语言中的 Boolean 类型中的 True 和 False。

7. sfr:特殊功能寄存器型

sfr 也是一种扩充数据类型,点用一个内存单元,值域为 0~255。利用它可以访问 51 单片机内部的所有特殊功能寄存器。如用 sfr P1=0x90 这一句定义 P1 为 P1 端口片内寄存器,在后面的语句中我们用以用 P1=255(对 P1 端口的所有引脚置高电平)之类的语句来操作特殊功能寄存器。

8. sfr16:16 位特殊功能寄存器类型

sfr16 占用两个内存单元,值域为 0~65535。sfr16 和 sfr 一样用于操作特殊功能寄存器,所不同的是它用于操作占两个字节的寄存器,如定时器 T0 和 T1。

9. sbit:可寻址位类型

sbit 是 C51 中的一种扩充数据类型,利用它可以访问芯片内部的 RAM 中的可寻址位或特殊功能寄存器中的可寻址位。如先前定义了 sfr P1=0x90;因 P1 端口的寄存器是可位寻址的,所以我们可以定义 sbit P1_1=P1^1;P1_1 为 P1 中的 P1.1 引脚。

同样可以用 P1.1 的地址去写,如 sbit P1_1=0x91;在以后的程序语句中就可以用 P1_1 来对 P1.1 引脚进行读写操作了。

通常这些可以直接使用系统提供的预处理文件,里面已定义好各特殊功能寄存器的简单名字,直接引用可以省去一点时间。当然您也可以自己写自己的定义文件,用您认为好记的名字。

sbit 的用法有三种:

第一种方法:sbit 位变量名=地址值;

第二种方法:sbit 位变量名=SFR 名称^变量位地址值;

第三种方法:sbit 位变量名=SFR 地址值^变量位地址值。

如定义 PSW 中的 OV 可以用以下三种方法:

sbit OV = 0xd2;   //0xd2 是 OV 的位地址值
sbit OV = PSW^2;   //其中 PSW 必须先用 sfr 定义好
sbit OV = 0xD0^2;   //0xD0 就是 PSW 的地址值

因此用 sfr P1_0=P1^0;就是定义用符号 P1_0 来表示 P1.0 引脚,也可以起 P10 一类的名字,只要下面程序中也随之更改就行了。

由于单片机体积小,不可能像微型计算机那样有很大的存储器。因此,首先必须根据需要指定各种变量的存放位置,这一点是 C51 与标准 C 语言的主要差异。C51 定义的存储器类型关键字及对应存储器变量的存储空间如表 4-2 所示。

表 4-2 C51 定义的存储器类型与存储空间

存储器类型	存储器空间	说 明
data	内部 RAM(00H—7FH)	128 字节,可直接寻址
bdata	内部 RAM(20H—2FH)	16 字节,可位寻址
Idata	内部 RAM(00H—FFH)	256 字节,间接寻址全部内部 RAM
pdata	外部 RAM(00H—FFH)	256 字节,用 MOVX @Ri 指令访问
xdata	外部 RAM(0000H—FFFFH)	64KB,用 MOVX @DPTR 指令访问
code	程序存储器(00H—FFFFH)	64KB,用 MOVC @A+DPTR 指令访问

三、C51 运算符与表达式

C 语言对数据有很强的表达能力,具有丰富的运算符。以下为 C51 中常用的运算符。

1. 算术运算符

① +　加或取正运算符

② -　减或取负运算符

③ *　乘运算符

④ /　除运算符

⑤ %　模运算(求余)。例如9%5结果是9除以5,得余数4。

算术运算符的优先级定为:先乘、除、模,后加、减,括号最优先。

2. 关系运算符

① <　小于运算符

② >　大于运算符

③ <=　小于等于运算符

④ >=　大于等于运算符

⑤ ==　等于运算符

⑥ !=　不等于运算符

关系符运算优先级①～④的优先级是相同的,后⑤～⑥也相同,但优先级要低于①～④。

关系表达式是二目运算符,它的作用在运算对象上产生一个逻辑值,或真或假。C语言来表示,1代表为真,0代表为假。

关系运算符的结合性为左结合,建议大家在用的时候用上括号,这样能更方便和不出错。

3. 逻辑运行符

① &&　逻辑"与"(AND)

② ||　逻辑"或"(OR)

③ !　逻辑"非"(NOT)

"&&"和"||"是双目运算,需要有两个对象,而"!"是单目运算,只要求有一个运行对象。

逻辑表达式和关系表达式一样,也会产生一个逻辑量真和假。以0代表为假,1代表为真。

4. 位运算符

① &　按位与

② |　按位或

③ ^　按位异或

④ ~　按位取反

⑤ <<　位左移

⑥ >>　位右移

除了按位取反"~"以外,所有的位操作全是两目运算符,要求有两个运算对象。位运算符只能是整型或字符型,不能为实型数据(浮点型)。

5. 自增减运算符

在C语言中也提供了自增减运算符,自增减运算符的作用是使变量值自动加1或减1。

++i,--i;　　　　　　　　//在使用i之前,先使i值加1或减1。

i++,i--;　　　　　　　　//在使用i之后,再使i值加1或减1。

注意:

自增减运算++、--只适合变量,而不能用于常量表达式。++、--的结合方式是自右向左的。

6. 赋值运算符

C51 的赋值运算符为"＝"，它的作用是将运算符右边的数据或表达式的值赋给运算符左边一个变量。赋值表达式的格式为：

变量＝表达式

例如：

a＝b＝0x32;   //将常数 0x32 同时赋值给变量 a,b

7. 复合赋值运算符

只要是二目运算，都可以与赋值运算符"＝"一起组成复合赋值运算符，C51 提供了 10 种复合赋值运算符。

＋＝，－＝，*＝，/＝，％＝，<<＝，>>＝，&＝，^＝，|＝。

例如：

A＋＝B 相当于 A＝A＋B；A－＝B 相当于 A＝A－B。

采用这种复合赋值运算的目的，是为了简化程序，提高 C 编译器效率。

8. 条件运算符

条件运算符的格式为：

逻辑表达式? 表达式 1:表达式 2

其功能是首先计算逻辑表达式，当值为真(非 0)时，将表达式 1 的值作为整个条件表达式的值；当值为假(0)时，将表达式 2 的值作为整个条件表达式的值。

例如：

max＝(a>b)? a:b 的执行结果是比较 a 与 b 的大小，若 a>b，则为真，max＝a；若 a<b，则为假，max＝b。

9. 指针和地址运算符

C51 的指针和地址运算符为：

* 取内容运算符

& 取地址运算符

取内容和地址的运算格式分别为：

变量 ＝ *指针变量;   //将指针变量所指向的目标变量值赋给左边变量
指针变量 ＝ &目标变量;   //将目标变量的地址赋给左边变量

例如：

px＝&i;   //将变量 i 的地址赋给指针 px
py＝*j;   //将 j 变量的内容为地址的单元的内容赋给 py

四、C51 结构化程序设计

C51 程序通常有如下三种控制结构：顺序结构、选择结构、循环结构。对于具体程序来说，每种结构都包含若干语句。C 语句可以分为空语句、表达式语句、函数调用语句、控制语句和复合语句 5 大类。

1. 空语句

只有一个分号的语句称为空语句,例如:

;

空语句什么也不做。C51 定义一个空函数语句 nop,用头文件 intrins.h 包含起来,然后在需要空语句的时候调用 nop 函数即可。

```
#include<intrins.h>
int nop( );
void main( )
{
int nop( );
}
```

2. 表达式语句

在一个表达式后面加上一个";",便构成一条表达式语句。例如:i=i+1;

3. 函数调用语句

由一个函数调用加一个分号便可构成函数调用语句。如:printf("This is a Cprogram.");

4. 控制语句

用于完成一定的控制功能,共有 9 种控制语句。

(1)选择结构控制语句

在 C51 中,选择结构由 if 语句或 switch 语句来实现。

① if 结构

格式:

if(表达式) 语句;

表示若表达式的值为非 0,则执行语句,否则跳过该语句执行后面的语句。

② if—else 结构

格式:

if(表达式) 语句 1;
else   语句 2;

上述结构表示:如果 if 后面表达式的值为非 0(即真),则执行语句 1,执行完语句 1 后跳过语句 2 开始继续向下执行;如果表达式的值为 0(即假),则跳过语句 1 而执行语句 2。

注意:

if 语句可以嵌套,这种情况经常碰到,但条件嵌套语句容易出错,其原因主要是不知道哪个 if 对应哪个 else。例如:

```
if(x>20||x<-10)
if(y<=100&&y>x)   printf("Good");
else              printf("Bad");
```

对于上述情况,规定:else 语句与最近的一个 if 语句匹配,上例中的 else 与 if(y<=100&&y>x) 相匹配。为了使 else 与 if(x>20||x<-10) 相匹配,必须用花括号。如下

所示:

```
if(x>20||x<-10)    { if(y<=100&&y>x)    printf("Good");}
else               printf("Bad");
```

③ if—else if 结构

格式:

```
if(表达式 1)         语句 1;
else if(表达式 2)    语句 2;
else if(表达式 3)    语句 3;
...
else                语句 n;
```

这种结构是从上到下逐个对条件进行判断,一旦发现条件满足就执行与它有关的语句,并跳过其他剩余阶梯;若没有一个条件满足,则执行最后一个 else 语句 n。最后这个 else 常起着"缺省条件"的作用。

④ switch 结构

在编写程序时,经常会碰到按不同情况分转的多路问题,这时可用嵌套 if—else—if 语句来实现,但 if—else—if 语句使用不方便,并且容易出错。对于这种情况,可以运用 switch 开关语句。开关语句格式为:

```
switch(变量)
{
    case 常量 1:语句 1;
    case 常量 2:语句 2;
    ...
    case 常量 n:语句 n;
    default:    语句 n+1;
}
```

执行 switch 开关语句时,将变量逐个与 case 后的常量进行比较,若与其中一个相等,则执行该常量下的语句,若不与任何一个常量相等,则执行 default 后面的语句。

(2)循环结构控制语句

三种基本的循环语句:for 语句、while 语句和 do—while 语句。

① for 循环语句

for 循环是开界的。它的一般形式为:

for(<初始化>;<条件表达式>;<增量>)

语句;

初始化总是一个赋值语句,它用来给循环控制变量赋初值;条件表达式是一个关系表达式,它决定什么时候退出循环;增量定义循环控制变量每循环一次后按什么方式变化。这三个部分之间用";"分开。

例如:

for(i=1;i<=10;i++)   语句;

上例中先给 i 赋初值 1，判断 i 是否小于等于 10，若是则执行语句，之后 i 值增加 1。再重新判断，直到条件为假，即 i>10 时，结束循环。

注意：

(1) for 循环中语句可以为语句体，但要用"{ }"将参加循环的语句括起来。

(2) for 循环中的"初始化"、"条件表达式"和"增量"都是选择项，即可以缺省，但";"不能缺省。省略了初始化，表示不对循环控制变量赋初值。省略了条件表达式，则不做其他处理时便成为死循环。省略了增量，则不对循环控制变量进行操作，这时可在语句体中加入修改循环控制变量的语句。

(3) for 循环可以有多层嵌套。

② while 循环语句

while 循环的一般形式为：

while(条件) 语句；

while 循环表示当条件为真时，便执行语句。直到条件为假才结束循环。并继续执行循环程序外的后续语句。

③ do－while 循环语句

do－while 循环的一般格式为：

do

语句；

while(条件)；

这个循环与 while 循环的不同在于：它先执行循环中的语句，然后再判断条件是否为真，如果为真则继续循环；如果为假，则终止循环。因此，do－while 循环至少要执行一次循环语句。

同样，当有许多语句参加循环时，要用"{ }"把它们括起来。

(3) break、continue 和 goto 语句

① break 语句

break 语句通常用在循环语句和开关语句中。当 break 用于开关语句 switch 中时，可使程序跳出 switch 而执行 switch 以后的语句；如果没有 break 语句，则将成为一个死循环而无法退出。

当 break 语句用于 do－while、for、while 循环语句中时，可使程序终止循环而执行循环后面的语句，通常 break 语句总是与 if 语句联在一起，即满足条件时便跳出循环。

注意：

(1) break 语句对 if－else 的条件语句不起作用。

(2) 在多层循环中，一个 break 语句只向外跳一层。

② continue 语句

continue 语句的作用是跳过循环体中剩余的语句而强行执行下一次循环。

continue 语句只用在 for、while、do－while 等循环体中，常与 if 条件语句一起使用，用来加速循环。

例如：

```
main( )
{
    char c;
    while(c! = 0X0D)                    //不是回车符则循环
    {
        c = getch();
        if(c = = 0X1B)
        continue;                        //若按 Esc 键不输出便进行下次循环
        printf("%c\n",c);
    }
}
```

③ goto 语句

goto 语句是一种无条件转移语句,与 BASIC 中的 goto 语句相似。goto 语句的使用格式为:

goto 标号;

其中标号是一个有效的标识符,这个标识符加上一个":"一起出现在函数内某处,执行 goto 语句后,程序将跳转到该标号处并执行其后的语句。另外标号必须与 goto 语句同处于一个函数中,但可以不在一个循环层中。通常 goto 语句与 if 条件语句连用,当满足某一条件时,程序跳到标号处运行。

goto 语句通常不用,主要因为大量使用该语句将使程序结构复杂,使层次变得不清楚,降低了程序的可理解性和可维护性。但在多层嵌套退出时,用 goto 语句则比较合理。

(4)return 语句

从函数返回。

5. 复合语句

用{ }括起来的语句称为复合语句。每条语句都以分号结束,如果只有一条语句,则{ }可以省略。

五、C51 函数

1. C51 函数的分类与定义

从用户使用的角度划分,C51 的函数分为两种:标准库函数和用户自定义函数。

(1)标准库函数

标准库函数是由 C51 编译器提供的,它不需要用户进行定义和编写,可以直接由用户调用。要使用这些标准库函数,必须在程序的开头用 #include 包含语句将指定文件的全部内容包含进来,然后才能调用。

C51 常用头文件如下:

absacc. h——包含允许直接访问 8051 不同存储区的宏定义;

assert. h——文件定义 assert 宏,可以用来建立程序的测试条件;

ctype——字符转换和分类程序;

intrins. h——文件包含指示编译器产生嵌入式固有代码的程序的原型;

math. h——数学程序;

reg51.h——51 的特殊寄存器；

reg52.h——52 的特殊寄存器；

setjmp.h——定义 jmp_buf 类型和 setjmp 和 longjmp 程序的原型；

stdarg.h——可变长度参数列表程序；

stdlib.h——存储区分配程序；

stdio.h——流输入和输出程序；

string.h——字符串操作程序、缓冲区操作程序。

打开 reg51.h 可以看到这样的一些内容：

```
/*--------------------------------------------------------------REG51.H
Header file for generic80C51 and 80C31 microcontroller.
Copyright (c) 1988-2001 Keil Elektronik GmbH and Keil Software,Inc. All rights reserved.
-----------------------------------------------------------------------*/

/* BYTE Register */
    sfr P0   = 0x80;          //P0 口
    sfr P1   = 0x90;          //P1 口
    sfr P2   = 0xA0;          //P2 口
    sfr P3   = 0xB0;          //P3 口
    sfr PSW  = 0xD0;          //程序状态字
    sfr ACC  = 0xE0;          //累加器
    sfr B    = 0xF0;          //寄存器,主要用于乘除
    sfr SP   = 0x81;          //堆栈指针,初始化为 07;先加 1 后压栈,先出栈再减 1
    sfr DPL  = 0x82;          //数据指针
    sfr DPH  = 0x83;          //数据指针
    sfr PCON = 0x87;          //电源控制
    sfr TCON = 0x88;          //Timer/Counter 控制
    sfr TMOD = 0x89;          //Timer/Counter 方式控制
    sfr TL0  = 0x8A;
    sfr TL1  = 0x8B;
    sfr TH0  = 0x8C;
    sfr TH1  = 0x8D;
    sfr IE   = 0xA8;          //中断控制
    sfr IP   = 0xB8;          //中断优先级控制
    sfr SCON = 0x98;          //串口控制寄存器
    sfr SBUF = 0x99;          //串口缓冲寄存器
    /* BIT Register */
    /* PSW */
    sbit CY  = 0xD7;          //进位或借位
    sbit AC  = 0xD6;          //辅助进借位
    sbit F0  = 0xD5;
    sbit RS1 = 0xD4;
    sbit RS0 = 0xD3;
    sbit OV  = 0xD2;
```

```
sbit P = 0xD0;
/* TCONRegister */
sbit TF1 = 0x8F;        //T1 的中断请求标志
sbit TR1 = 0x8E;        //Timer 1 running
sbit TF0 = 0x8D;        //T0 的中断请求标志
sbit TR0 = 0x8C;        //Timer 0 running
sbit IE1 = 0x8B;        //interrupt external 1 外中断请求标志
sbit IT1 = 0x8A;        //interrupt triggle 1 外中断触发方式
sbit IE0 = 0x89;        //interrupt external 0 外中断请求标志
sbit IT0 = 0x88;        //interrupt triggle 0 外中断触发方式
/* IERegister */
sbit EA = 0xAF;         //开串口中断
sbit ES = 0xAC;
sbit ET1 = 0xAB;
sbit EX1 = 0xAA;
sbit ET0 = 0xA9;
sbit EX0 = 0xA8;
/* IPRegister */
sbit PS = 0xBC;         //串行中断优先级
sbit PT1 = 0xBB;        //T1 优先级
sbit PX1 = 0xBA;        //外部中断 1 优先级
sbit PT0 = 0xB9;        //T0 优先级
sbit PX0 = 0xB8;        //外部中断 0 优先级
/* P3Register */
sbit RD = 0xB7;
sbit WR = 0xB6;
sbit T1 = 0xB5;
sbit T0 = 0xB4;
sbit INT1 = 0xB3;       //外中断 1
sbit INT0 = 0xB2;       //外中断 0
sbit TXD = 0xB1;        //串行发送
sbit RXD = 0xB0;        //串行接收
/* SCONRegister */
sbit SM0 = 0x9F;
sbit SM1 = 0x9E;        //串口工作方式
sbit SM2 = 0x9D;
sbit REN = 0x9C;        //串行接收允许
sbit TB8 = 0x9B;        //收到的第九位
sbit RB8 = 0x9A;        //要发的第九位
sbit TI = 0x99;         //接收完成中断标志
sbit RI = 0x98;         //接收完成中断标志
```

(2)用户自定义函数

用户自定义函数是用户根据自己的需要编写的能实现特定功能的函数,它必须先进行定义才能调用。函数定义的一般形式为:

函数类型 函数名(形式参数表)
{
    局部变量定义
    函数体语句
}

2. C51 函数的说明与调用

在调用一个函数之前,必须对该函数的类型进行说明。对函数进行说明的一般形式为:

类型标识符 被调用的函数名(形式参数表);

函数说明与函数定义是不同的,书写时必须注意函数说明结束时,必须加上一个分号";"。如果被调用函数在主调用函数之前已经定义了,则不需要进行说明;否则需要在主调用函数前对被调用函数进行说明。

C51 程序中的函数是可以互相调用的。函数调用的一般形式为:

函数名(实际参数表)

其中,"函数名"就是被调用的函数,"实际参数表"就是与形式参数表对应的一组变量,它的作用就是将实际参数的值传递给被调用函数中的形式参数。在调用时,实际参数与形式参数必须在个数、类型、顺序上严格一致。

函数的调用有以下三种:

(1)函数语句。如:fun(a,b);
(2)函数表达式。如:result=5*fun(a,b);
(3)函数实参。如:result=fun(fun(a,b),c)。

3. C51 中的特殊函数

(1)再入函数

如果在调用一个函数的过程中,又间接或直接调用该函数本身,称为函数的递归调用。在 C51 中必须采用一个扩展关键字 reentrant 作为定义函数时的选项,将该函数定义为再入函数,此时该函数才可被递归调用。

再入函数的定义格式为:

函数类型 函数名(形式参数表)[reentrant]

由于采用再入函数时需要用再入栈来保存相关变量数据,占用较大内存,处理速度较慢,因此,一般情况下尽量避免使用递归调用。

(2)中断服务函数

C51 编译器支持在 C 语言源程序中直接编写 51 单片机的中断服务函数程序。以前学习用汇编语言编写中断服务程序时,会对堆栈出栈的保护问题而觉得头痛。为了能够在 C 语言源程序中直接编写中断服务函数。C51 编译器对函数的定义进行了扩展。增加了一个扩展关键字 interrupt。关键字 interrupt 是函数定义时的一个选项,加上这个选项就可以将一个函数定义成中断服务函数。定义中断服务函数的一般形式为：

函数类型 函数名（形式参数表）［interrupt m］［using n］

关键字 interrupt 后面的 m 是中断号，取值范围为 $0\sim31$。编译器从 $8m+3$ 处产生中断向量。具体的中断号 m 和中断向量取决于不同的 51 系列单片机芯片。51 单片机常用中断源和中断向量如表 4-3 所示。

表 4-3 常用的中断源和中断向量表

中断编号 m	中断源	中断入口地址（中断向量 $8m+3$）
0	外部中断 0	0003H
1	定时器/计数器 0 溢出中断	000BH
2	外部中断 1	0013H
3	定时器/计数器 1 溢出中断	001BH
4	串行口中断	0023H

51 系列单片机可以在内部 RAM 中使用 4 个不同的工作寄存器组，每个寄存器组中包含 8 个工作寄存器（R0～R7）。C51 编译器扩展了一个关键字 using，专门用来选择 51 单片机中不同的工作寄存器组。using 后面的 n 是一个 0～3 的常整数，分别选中 4 个不同的工作寄存器组。在定义一个函数时 using 是一个选项。如果不用该选项，则由编译器选择一个寄存器组作绝对寄存器组访问。

关键字 using 对函数目标代码的影响如下：

在函数的入口处将当前工作寄存器组保护到堆栈中，指定的工作寄存器内容不会改变，函数返回之前将被保护的工作寄存器组从堆栈中恢复。使用关键字 using 在函数中确定一个工作寄存器组时必须十分小心，要保证任何寄存器组的切换都只在控制的区域内发生。如果不做到这一点将产生不正确的函数结果。

另外，带 using 属性的函数，原则上不能返回 bit 类型的值，并且关键字 using 不允许用于外部函数，关键字 interrupt 也不允许用于外部函数，它对中断函数目标代码的影响如下：

在进入中断函数时，特殊功能寄存器 ACC、B、DPH、DPL、PSW 将被保存入栈。如果不使用寄存组切换，则将中断函数中所用到的全部工作寄存器都入栈。函数返回之前，所有的寄存器内容出栈。中断函数由 51 单片机指令 RETI 结束。

值得注意的是，编写 51 单片机中断函数时应严格遵循以下规则：

① 中断函数不能进行参数传递，如果中断函数中包含任何参数声明都将导致编译出错。

② 中断函数没有返回值，如果企图定义一个返回值将得到不正确的结果。因此最好将中断函数定义为 void 类型，以明确说明没有返回值。中断服务函数最好写在文件的尾部，并且禁止使用 extern 存储种类说明。

③ 在任何情况下都不能直接调用中断函数，否则会产生编译错误。因为中断函数的返回是 51 单片机指令 RETI 完成的，RETI 指令影响 51 单片机的硬件中断系统。

④ 如果中断函数中用到浮点运算，必须保存浮点寄存器的状态。当没有其他程序执行

浮点运算时可以不保存。

⑤ 如果在中断函数中调用了其他函数,则被调用函数所使用的寄存器组须与中断函数相同。用户必须保证按要求使用相同的寄存器组,否则会产生不正确的结果。如果定义中断函数时没有使用 using 选项,则由编译器选择一个寄存器组作绝对寄存器组访问。

4.C51 中直接嵌入汇编

为了发挥 C51 语言和汇编语言各自的优点,常需要将两者进行混合编程。一般情况下,由于 C51 具有很强的数据处理能力,编程中对 51 单片机寄存器和存储器的分配由编译器自动完成,因此常用它来编写主程序及一些运算教复杂的程序。而汇编语言对硬件的控制较强,运行速度快,灵活性强,因此常用汇编语言实现与硬件接口的子程序设计以及对时间要求高的子程序设计。下面简单介绍 C51 调用汇编程序的方法。

(1)在 C 文件中要嵌入汇编代码片以如下方式加入汇编代码:

```
#pragma ASM
;Assembler Code Here
#pragma ENDASM
```

(2)在 Project 窗口中包含汇编代码的 C 文件上右键,选择"Options for...",点击右边的"Generate Assembler SRC File"和"Assemble SRC File",使检查框由灰色变成黑色(有效)状态。

(3)根据选择的编译模式,把相应的库文件(如 Small 模式时,是 Keil\C51\Lib\C51S.Lib)加入工程中,该文件必须作为工程的最后文件。

(4)编译,即可生成目标代码。

六、C51 数组

在程序设计中,为了处理方便,把具有相同数据类型的若干变量按有序的形式组织起来。这些按序排列的同类数据元素的集合称为数组。形象的能这样去理解,就像一个学校在操场上排队,每一个年级代表一个数据类型,每一个班级为一个数组,每一个学生就是数组中的一个数据。数据中的每个数据都能用唯一的下标来确定其位置,下标能是一维或多维的。就如在学校的方队中要找一个学生,这个学生在 I 年级 H 班 X 组 Y 号,那么能把这个学生看做在 I 类型的 H 数组中(X,Y)下标位置中。数组和普通变量一样,要求先定义了才能使用,下面是定义一维或多维数组的方式:

数据类型说明符 数组名[常量表达式];
数据类型说明符 数组名[常量表达式 1]……[常量表达式 N];

"数据类型说明符"是指数组中的各数据单元的类型,每个数组中的数据单元只能是同一数据类型。

"数组名"是整个数组的标识,命名方法和变量命名方法是一样的。在编译时系统会根据数组大小和类型为变量分配空间,数组名就是所分配空间的首地址的标识。

"常量表达式"是表示数组的长度和维数,它必须用"[]"括起,括号里的数不能是变量只能是常量。

```
unsigned int xcount[10];        //定义无符号整形数组,有 10 个数据单元
char inputstring[5];            //定义字符形数组,有 5 个数据单元
```

任务4 C51程序设计知识

```
float outnum[10][10];           //定义浮点型数组,有100个数据单元
```

在C语言中数组的下标是从0开始的而不是从1开始,如一个具有10个数据单元的数count,它的下标就是从count[0]到count[9],引用单个元素就是数组名加下标,如count[1]就是引用count数组中的第2个元素,如果错用count[10]就会有错误出现了。还有一点要注意的就是在程序中只能逐个引用数组中的元素,不能一次引用整个数组,但是字符型的数组就能一次引用整个数组。

在上面介绍的定义方式只适用于定义在内存DATA存储器使用的内存,有时候我们需要把一些数据表存放在数组中,通常这些数据是不用在程序中改变数值的,这个时候就要把这些数据在程序编写时就赋给数组变量。因为51芯片的片内RAM很有限,通常会把RAM分给参与运算的变量或数组,而那些程序中不变数据则应存放在片内的CODE存储区,以节省宝贵的RAM。

数组在使用前需要赋初值。赋初值的方式如下:

数据类型 [存储器类型] 数组名[常量表达式] = {常量表达式};
数据类型 [存储器类型] 数组名[常量表达式1]......[常量表达式N] = {{常量表达式1}......{常量表达式N}};

在定义并为数组赋初值时,初学者一般会搞错初值个数和数组长度的关系,而致使编译出错。初值个数必须小于或等于数组长度,不指定数组长度则会在编译时由实际的初值个数自动设置。

```
unsigned char LEDNUM[2] = {12,35};      //一维数组赋初值
int Key[2][3] = {{1,2,4},{2,2,1}};       //二维数组赋初值
unsigned char IOStr[] = {3,5,2,5,3};     //没有指定数组长度,编译器自动设置
unsigned char code skydata[] = {0x02,0x34,0x22,0x32,0x21,0x12};//数据保存在code区
```

七、C51指针

指针(pointer)就是指变量或数据所在的存储区地址。如一个字符型的变量str存放在内存单元DATA区的51H这个地址中,那么DATA区的51H地址就是变量str的指针。在C语言中,指针是一个很重要的概念,正确有效的使用指针类型的数据,就能更有效的表达复杂的数据结构,能更有效的使用数组或变量,能方便直接的处理内存或其它存储区。指针之所以能这么有效的操作数据,是因为无论程序的指令、常量、变量或特殊寄存器都要存放在内存单元或相应的存储区中,这些存储区是按字节来划分的,每一个存储单元都能用唯一的编号去读或写数据,这个编号就是常说的存储单元的地址,而读写这个编号的动作就叫做寻址,通过寻址就能访问到存储区中的任一个能访问的单元,而这个功能是变量或数组等是不可能代替的。C语言也因此引入了指针类型的数据类型,专门用来确定其他类型数据的地址。

用一个变量来存放另一个变量的地址,那么用来存放变量地址的变量称为"指针变量"。如用变量strip来存放str变量的地址51H,变量strip就是指针变量。下面用图表4-1来说明变量的指针和指针变量两个不一样的概念。

变量的指针就是变量的地址,用取地址运算符"&"取得赋给指针变量。&str就是把变量str的地址取得。用语句strip = &str就能把所取得的str指针存放在strip指针变量

中。strip 的值就变为 51H。可见指针变量的内容是另一个变量的地址,地址所属的变量称为指针变量所指向的变量。

要访问变量 str 除了能用"str"这个变量名来访问之外,还能用变量地址来访问。方法是先用 &str 取变量地址并赋予 strip 指针变量,然后就能用 *strip 来对 str 进行访问了。

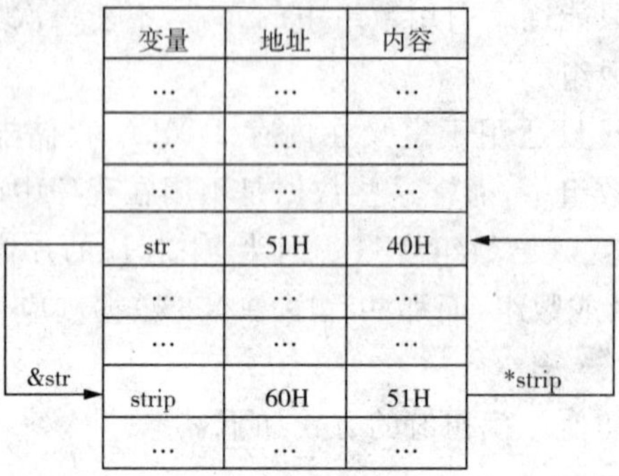

变量	地址	内容
...	...	...
...	...	...
...	...	...
str	51H	40H
...	...	...
...	...	...
strip	60H	51H
...	...	...

图 4-1 变量的指针和指针变量示意图

"*"是指针运算符,用它能取得指针变量所指向的地址的值。在上图中指针变量 strip 所指向的地址是 51H,而 51H 中的值是 40H,那么 *strip 所得的值就是 40H。使用指针变量之前也和使用其他类型的变量那样要求先定义变量,而且形式也相类似。指针一般的定义形式如下:

数据类型　　[存储器类型]　　*变量名;
unsigned　char　xdata　*pi;
//指针会占用二字节,指针自身存放在编译器默认存储区,指向 xdata 存储区的 char 类型
unsigned char xdata　* data pi;　　　//指针指定在 data 区,其他同上
int * pi;
//定义为一般指针,指针自身存放在编译器默认存储区,占三个字节

在定义形式中"数据类型"是指所定义的指针变量所指向的变量的类型。"存储器类型"是编译器编译时的一种扩展标识,它是可选的。在没有"存储器类型"选项时,则定义为一般指针,如有"存储器类型"选项时则定义为基于存储器的指针。限于 51 芯片的寻址范围,指针变量最大的值为 0xFFFF,这样就决定了一般指针在内存会占用 3 个字节,第一字节存放该指针存储器类型编码,后两个则存放该指针的高低位址。而基于存储器的指针因为不用识别存储器类型所以会占一或二个字节,idata、data、pdata 存储器指针占一个字节,code、xdata 则会占二个字节。

由上可知,明确的定义指针,能节省存储器的开销,这在严格要求程序体积的项目中很有用处。

任务 5　闪烁灯控制

实训任务

在单片机的 P0.0 口线上接一个发光二极管 L1,使 L1 不停地一亮一灭,时间间隔为 0.2 秒。

实训设备

1. 设备
PC 机(安装 wave 编程软件、Keil C51 软件)、单片机实验板。
2. 工具及材料
工作对象:电工电子工具、电子元器件和辅助材料、仿真器、编程器。
工作工具:单片机控制电路原理图、实训指导书、项目任务单、工作记录单、项目检查单、各种电工仪表、常用电工工具和拆装工具、量具、相关电子手册。

硬件设计

主控模块采用 ATMEL 公司生产的 AT89S52 单片机。

电源、时钟信号以及复位电路是单片机工作的基本条件,缺一不可。单片机系统的基本工作电路包括电源电路、时钟电路、复位电路。

LED 信号灯电路采用发光二极管 LED,LED 的 K 极通过限流电阻 R 与单片机 P0 口的 P0.0 引脚连接,LED 的 A 极连接到+5V 电源(系统供电电源为+5V),LED 上串接的电阻是 1kΩ,如果此时 LED 上的电压是 2.0V,那么此时通过 LED 的电流为(5V～2V)/1000Ω=3mA。如果需要提高亮度,电流一般会控制在 10mA 左右,则此时电阻应该选择(5V～2V)/10mA=300Ω,实际中可以近似选择 330Ω。LED 模块与单片机的接口电路如图 5-1 所示。

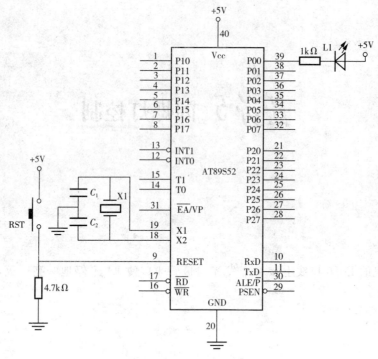

图 5-1 LED闪烁灯模块与单片机的接口电路图

软件设计

一、延时程序的设计方法

由于单片机的指令的执行时间很短,数量为微秒级。因此,如果要求的闪烁时间间隔为0.2秒,相对于微秒相差太大,所以在执行某一指令时,插入延时程序,来达到要求,但这样的延时程序是如何设计呢?

延时子程序用于提供发光二极管 L1 点亮或熄灭的延时时间。

延时程序是一种应用较为广泛的小程序,一般采用多条语句循环执行来实现延时。下面具体介绍其原理。

系统所用的石英晶体振荡器的频率为 11.0592MHz,因此,1 个机器周期=1/石英频率×12,即为 12/11.0592us。

```
DELAY200ms:MOV R5,#20      ;2个机器周期
LOOP1:MOV R6,#20           ;2个机器周期
LOOP2:MOV R7,#230          ;2个机器周期
DJNZ R7,$                  ;2个机器周期
DJNZ R6,LOOP2              ;2个机器周期
DJNZ R5,LOOP1              ;2个机器周期
```

执行上面的程序需要的时间大约为:R5×R6×R7×2×机器周期=20×20×230×2×12/11.0592=199652.78us=199.65278ms=0.19965278s,即大约为 0.2s。

二、算法设计

根据系统工作原理图 5-1 可知：

当 P0.0 端口输出高电平，即 P0.0＝1 时，根据发光二极管的单向导电性可知，这时发光二极管 L1 熄灭，我们可以使用"L1＝1"指令使 P0.0 端口输出高电平。

当 P0.0 端口输出低电平，即 P0.0＝0 时，发光二极管 L1 亮，可以使用"L1＝0"指令使 P0.0 端口输出低电平。

1. 延时函数设计

由于单片机指令的执行时间很短，属于微秒级。而要求的彩灯闪烁时间间隔为 0.2s，相对于微秒相差太大，所以在执行彩灯点亮和熄灭指令时，应插入延时程序来达到我们的要求。

由于发光二极管点亮的时间为 0.2s，熄灭的时间也为 0.2s，在程序中需要多次执行同样的计算和操作，如果每次都从头开始编制该段程序，不仅麻烦，而且浪费存储空间。对于这种在一个程序中反复出现的程序段，我们采用函数模块来实现。

2. 循环程序结构设计

由于需要控制发光二极管反复的一亮一灭，在程序中需要反复执行该程序段，为了避免在程序中多次的编写，我们采用循环结构来实现该功能，可以通过利用循环结构 while(1) 来控制程序的反复执行，以实现发光二极管反复的一亮一灭。

三、程序设计

1. 主函数设计

主函数主要完成硬件初始化、函数调用等功能。

(1) 初始化

把 P0.0 端口初始化为 0，让发光二极管 L1 点亮。

(2) 循环闪烁

首先调用延时函数 delay02s()，使发光二极管 L1 点亮 0.2s；然后修改 P0.0 的状态为 1，再调用延时函数 delay02s()，使发光二极管 L1 熄灭 0.2s；最后使程序跳转到开始再重新循环执行，就可以实现发光二极管的循环闪烁。

主函数设计流程图如图 5-2 所示。

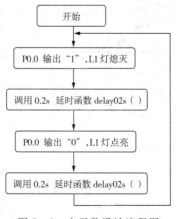

图 5-2 主函数设计流程图

2. 延时函数设计

延时函数用于提供发光二极管 L1 点亮或熄灭的延时时间。系统所用的石英晶体振荡器的频率为 11.0592MHz，因此，1 个机器周期＝1/石英频率×12，即为 12/11.0592us。

```
void delay02s(void)
{
    unsigned char i,j,k;            //定义无符号字符型变量i,j、k
    for(i=20;i>0;i--)               //外层循环
        for(j=20;j>0;j--)           //中层循环
            for(k=230;k>0;k--);     //内层循环
}
```

执行上面的程序需要的时间大约为：i×j×k×2×机器周期＝20×20×230×2×12/11.0592＝199652.78us＝199.65278ms＝0.19965278s，即大约为 0.2s。

延时函数设计流程图如图 5-3 所示。

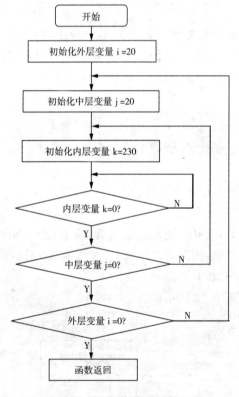

图 5-3　延时函数设计流程图

四、汇编语言源程序

```
;*************************************************************
;项目名称:闪烁灯控制
;功能:发光二极管不停地一亮一灭,时间间隔为 0.2s,循环往复
;*************************************************************
;主程序
```

任务 5　闪烁灯控制

```
            ORG 0000H              ;程序入口地址
START:CLR  P0.0                    ;当 P0.0=0 时,发光二极管 L1 点亮
            LCALL DELAY200ms       ;调用延时子程序,LED 灯点亮 200ms
            SETB  P0.0             ;当 P0.0=1 时,发光二极管 L1 熄灭
            LCALL DELAY200ms       ;调用延时子程序,L1 熄灭 200ms
            LJMP  START            ;跳转到程序开始,L1 一亮一灭,循环往复
;延时 0.2s 子程序 DELAY200ms
DELAY200ms:MOV R5,#20;              外层循环结构开始,设置外层循环变量初始值
LOOP1:       MOV R6,#20             ;中层循环结构开始,设置中层循环变量初始值
LOOP2:       MOV R7,#230            ;内层循环结构开始,设置内层循环变量初始值
             DJNZ R7,$              ;内层循环结束判断条件,直到 R7=0 为止
             DJNZ R6, LOOP2         ;中层循环结束判断条件,直到 R6=0 为止
             DJNZ R5, LOOP1         ;外层循环结束判断条件,直到 R5=0 为止
             RET                    ;延时 0.2s 子程序 DELAY200ms 返回
             END                    ;程序结束
```

五、C 语言源程序

```c
//****************************************************************
//项目名称:闪烁灯控制
//功能:发光二极管不停地一亮一灭,时间间隔为 0.2s,循环往复
//****************************************************************
#include<AT89X51.H>             //包含"51 寄存器定义"头文件
//变量定义
sbit L1=P0^0;                   //定义发光二极管 L1 为 P0.0
//函数声明:延时 0.2s 函数 delay02s()
void delay02s(void);
//主函数
void main(void)                 //主函数入口地址
{
    while(1)                    //进行下一轮发光二极管循环点亮
    {
        L1=0;                   //当 P0.0=0 时,发光二极管 L1 点亮
        delay02s();             //调用函数 delay02s()延时 0.2s
        L1=1;                   //当 P0.0=1 时,发光二极管 L1 熄灭
        delay02s();             //调用函数 delay02s()延时 0.2s
    }
}
//延时 0.2s 函数 delay02s()
void delay02s(void)
{
    unsigned char i,j,k;        //定义无符号字符型变量 i、j、k
    for(i=20;i>0;i--)           //外层循环
        for(j=20;j>0;j--)       //中层循环
```

```
        for(k=230;k>0;k--);      //内层循环
}
```

实训考核

本课程改革传统的闭卷或开卷考核,而采用过程考核为主的多元化考核方式,考核分为理论考核、职业道德考核和技能考核三部分,各部分所占比例见表5-1、表5-2、表5-3所示。

表5-1 理论考核和职业素质考核形式及所占比例

序号	名称			比例	得分
一	理论考核	过程作业文件	个人自评	10%	
			组内互评	20%	
			小组互评	30%	
			老师评定	20%	
		课堂提问、解答		10%	
		项目汇报		10%	
		小计		100%	
二	职业素质	职业道德,工作作风		40%	
		小组沟通协作能力		40%	
		创新能力		20%	
		小计		100%	

表5-2 技能考核内容及比例

姓名		班级		小组		总得分	
序号	考核项目	考核内容及要求		配分	评分标准	考核环节	得分
1	①安全文明生产 ②安全操作规范	着装规范		20%	现场考评	实施	
		安全用电					
		布线规范、合理					
		工具摆放整齐					
		工具及仪器仪表使用规范、摆放整齐					
		任务完成后,进行场地整理,保持场地清洁、有序					
2	实训态度	不迟到、早退、旷课		10%	现场考评	六步	
		实训过程认真负责					
		组内主动沟通、协作,小组间互助					

(续表)

3	系统方案制定	工作流程正确合理	10%	现场考评	计划决策	
		方案合理				
		选用指令是否合理				
		电路图正确				
	编程能力	独立完成程序	10%	现场考评	决策	
		程序简单、可靠				
4	操作能力	正确输入程序并进行程序调试	20%	现场考评	实施	
		根据电路图正确接线				
		根据系统功能进行正确操作演示				
5	工艺	接线美观	10%	现场考评	实施	
		线路工作可靠				
6	实践效果	系统工作可靠	10%	现场考评	检查	
		满足工作要求				
		创新				
		按规定的时间完成项目				
7	汇报总结	工作总结,PPT汇报	5%	现场考评	评估	
		填写自我检查表及反馈表				
8	技术文件制作整理	技术文件制作整理能力	5%	现场考评	评估	
合计			100%			

表5-3 各部分考核占课程考核的比例

考核项目	理论考核	技能考核	职业素质考核	合　计
比　例	30%	50%	20%	100%
分　值	30	50	20	100
实际得分				

任务6 模拟开关灯控制

实训任务

监视按钮开关 K1（接在 P1.0 端口上），用发光二极管 L1（接在单片机 P0.0 端口上）显示开关状态：如果 K1 按下，则 L1 亮；如果 K1 松开，则 L1 熄灭。

实训设备

1. 设备

PC 机（安装 wave 编程软件、Keil C51 软件）、单片机实验板。

2. 工具及材料

工作对象：电工电子工具、电子元器件和辅助材料、仿真器、编程器。

工作工具：单片机控制电路原理图、实训指导书、项目任务单、工作记录单、项目检查单、各种电工仪表、常用电工工具和拆装工具、量具、相关电子手册。

硬件设计

主控模块采用 ATMEL 公司生产的 AT89S52 单片机。模拟开关灯模块与单片机的接口电路如图 6-1 所示。

软件设计

一、设计思想

1. 按钮开关状态的检测过程

单片机对开关状态的检测相对于单片机来说，是从单片机的 P1.0 端口输入信号，而输入的信号只有高电平和低电平两种，当松开时，即输入高电平，相当开关断开，当按下时，即输入低电平，相当开关闭合。单片机可以采用 JB BIT,REL 或者是 JNB BIT,REL 指令来完成对开关状态的检测即可。

任务6 模拟开关灯控制

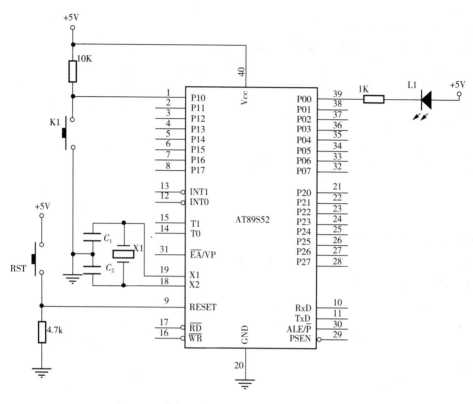

图6-1 模拟开关灯模块与单片机的接口电路图

2．输出控制

如图6-2所示，当 P0.0 端口输出高电平，即 P0.0＝1 时，根据发光二极管的单向导电性可知，这时发光二极管 L1 熄灭；当 P0.0 端口输出低电平，即 P0.0＝0 时，发光二极管 L1 亮；我们可以使用 SETB P0.0 指令使 P0.0 端口输出高电平，使用 CLR P0.0 指令使 P0.0 端口输出低电平。

二、程序流程图

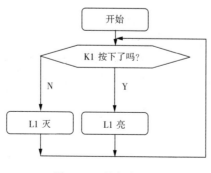

图6-2 程序流程图

三、汇编语言源程序

```
;*********************************************************************
;项目名称:模拟开关灯控制
;功能:监视按钮开关 K1 状态:如果 K1 按下,则 L1 点亮;如果 K1 松开,则 L1 熄灭。
;*********************************************************************
            ORG     0000H
    START:  JB      P1.0,LIG
            CLR     P0.0
            SJMP    START
    LIG:    SETB    P0.0
            SJMP    START
            END
```

四、C 语言源程序

```c
//*********************************************************************
//项目名称:模拟开关灯控制
//功能:监视按钮开关 K1 状态:如果 K1 按下,则 L1 点亮;如果 K1 松开,则 L1 熄灭。
//*********************************************************************
#include<AT89X51.H>
sbit K1 = P1^0;
sbit L1 = P0^0;
void main(void)                    //主函数
{
    while(1)
    {
      if(K1==0)   {L1=0;}          //K1 按下,灯 L1 点亮
      else        {L1=1;}          //K1 松开,灯 L1 熄灭
    }
}
```

实训考核

本课程改革传统的闭卷或开卷考核,而采用过程考核为主的多元化考核方式,考核分为理论考核、职业道德考核和技能考核三部分,各部分所占比见表 6-1、表 6-2、表 6-3。

表6-1 理论考核和职业素质考核形式及所占比例

序号	名称		比例	得分
一	理论考核	个人自评	10%	
		组内互评	20%	
	过程作业文件	小组互评	30%	
		老师评定	20%	
	课堂提问、解答		10%	
	项目汇报		10%	
	小计		100%	
二	职业素质	职业道德,工作作风	40%	
		小组沟通协作能力	40%	
		创新能力	20%	
	小计		100%	

表6-2 技能考核内容及比例

姓名		班级		小组		总得分	
序号	考核项目	考核内容及要求		配分	评分标准	考核环节	得分
1	①安全文明生产 ②安全操作规范	着装规范		20%	现场考评	实施	
		安全用电					
		布线规范、合理					
		工具摆放整齐					
		工具及仪器仪表使用规范、摆放整齐					
		任务完成后,进行场地整理,保持场地清洁、有序					
2	实训态度	不迟到、早退、旷课		10%	现场考评	六步	
		实训过程认真负责					
		组内主动沟通、协作,小组间互助					
3	系统方案制定	工作流程正确合理		10%	现场考评	计划决策	
		方案合理					
		选用指令是否合理					
		电路图正确					
	编程能力	独立完成程序		10%	现场考评	决策	
		程序简单、可靠					

(续表)

4	操作能力	正确输入程序并进行程序调试	20%	现场考评	实施
		根据电路图正确接线			
		根据系统功能进行正确操作演示			
5	工艺	接线美观	10%	现场考评	实施
		线路工作可靠			
6	实践效果	系统工作可靠	10%	现场考评	检查
		满足工作要求			
		创新			
		按规定的时间完成项目			
7	汇报总结	工作总结，PPT汇报	5%	现场考评	评估
		填写自我检查表及反馈表			
8	技术文件制作整理	技术文件制作整理能力	5%	现场考评	评估
合计			100%		

表6-3 各部分考核占课程考核的比例

考核项目	理论考核	技能考核	职业素质考核	合　计
比　例	30%	50%	20%	100%
分　值	30	50	20	100
实际得分				

任务 7 多路开关状态指示灯控制

实训任务

单片机的 P0.0～P0.3 接四个发光二极管 L1～L4,P1.0～P1.3 接四个按钮开关 K1～K4,编程将开关的状态反映到发光二极管上(若开关闭合,对应的灯亮;若开关断开,对应的灯灭)。

实训设备

1. 设备

PC 机(安装 wave 编程软件、Keil C51 软件)、单片机实验板。

2. 工具及材料

工作对象:电工电子工具、电子元器件和辅助材料、仿真器、编程器。

工作工具:单片机控制电路原理图、实训指导书、项目任务单、工作记录单、项目检查单、各种电工仪表、常用电工工具和拆装工具、量具、相关电子手册。

硬件设计

主控模块采用 ATMEL 公司生产的 AT89S52 单片机。多路开关状态指示模块与单片机的接口电路如图 7-1 所示。

软件设计

一、开关状态检测

对于开关状态检测,相对单片机来说,是输入关系,我们可轮流检测每个开关状态,根据每个开关的状态让相应的发光二极管指示,可以采用 JB P1.X,REL 或 JNB P1.X,REL 指令来完成;也可以一次性检测四路开关状态,然后让其指示,可以采用 MOV A,P1 指令一次把 P1 端口的状态全部读入,然后取低 4 位的状态来指示。

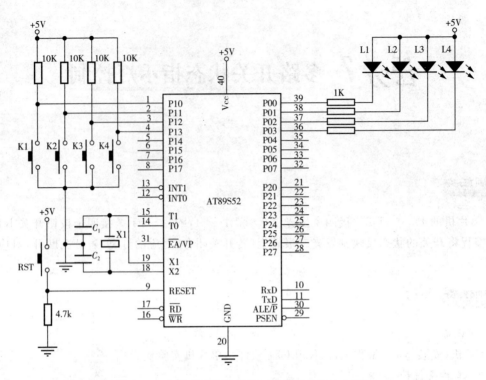

图 7-1 多路开关状态指示与单片机的接口电路图

二、输出控制

根据开关的状态,由发光二极管 L1~L4 来指示,可以用 SETB P0.X 和 CLR P0.X 指令来完成,也可以采用 MOV P0,#1111XXXXB 方法一次指示。

三、程序流程图

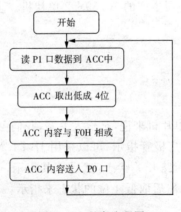

图 7-2 程序流程图

四、汇编语言源程序

```
;**************************************************************
;项目名称:多路开关状态指示灯控制
;功能:将开关的状态反映到 LED 上(若开关闭合,对应的灯亮;若开关断开,对应的灯灭)
;**************************************************************
;方法一
        ORG     0000H
START:  MOV     A,P1
        ORL     A,#0F0H
        MOV     P0,A
        SJMP    START
        END
;方法二
        ORG     0000H
START:  JB      P1.0,L1G
        CLR     P0.0
        SJMP    NEXT1
L1G:    SETB    P0.0
NEXT1:  JB      P1.1,L2G
        CLR     P0.1
        SJMP    NEXT2
L2G:    SETB    P0.1
NEXT2:  JB      P1.2,L3G
        CLR     P0.2
        SJMP    NEXT3
L3G:    SETB    P0.2
NEXT3:  JB      P1.3,L4G
        CLR     P0.3
        SJMP    START
L4G:    SETB    P0.3
        SJMP    START
        END
```

五、C 语言源程序

```
//**************************************************************
//项目名称:多路开关状态指示灯控制
//功能:将开关的状态反映到 LED 上(若开关闭合,对应的灯亮;若开关断开,对应的灯灭)
//**************************************************************
//方法一
#include<AT89X51.H>
unsigned char temp;
void main(void)
{
```

```
            while(1)
             {
                temp = P1;
                temp = temp|0xf0;
                P0 = temp;
             }
         }
//方法二
#include<AT89X51.H>
sbit K1 = P1^0;
sbit LED1 = P0^0;
sbit K2 = P1^1;
sbit LED2 = P0^1;
sbit K3 = P1^2;
sbit LED3 = P0^2;
sbit K4 = P1^3;
sbit LED4 = P0^3;
void main(void)                            //主函数
{
   while(1)
     {
            if(K1= =0)   {LED1=0;}      //灯1亮
            else         {LED1=1;}      //灯1灭
            if(K2= =0)   {LED2=0;}      //灯2亮
            else         {LED2=1;}      //灯2灭
            if(K3= =0)   {LED3=0;}      //灯3亮
            else         {LED3=1;}      //灯3灭
            if(K4= =0)   {LED4=0;}      //灯4亮
            else         {LED4=1;}      //灯4灭
     }
}
```

实训考核

本课程改革传统的闭卷或开卷考核,而采用过程考核为主的多元化考核方式,考核分为理论考核、职业道德考核和技能考核三部分,各部分所占比例见表7-1、表7-2、表7-3。

任务7 多路开关状态指示灯控制

表 7-1 理论考核和职业素质考核形式及所占比例

序 号	名 称		比 例	得 分
一	理论考核	过程作业文件 — 个人自评	10%	
		过程作业文件 — 组内互评	20%	
		过程作业文件 — 小组互评	30%	
		过程作业文件 — 老师评定	20%	
		课堂提问、解答	10%	
		项目汇报	10%	
		小计	100%	
二	职业素质	职业道德,工作作风	40%	
		小组沟通协作能力	40%	
		创新能力	20%	
		小计	100%	

表 7-2 技能考核内容及比例

姓名		班级		小组		总得分	
序号	考核项目	考核内容及要求		配分	评分标准	考核环节	得分
1	①安全文明生产 ②安全操作规范	着装规范		20%	现场考评	实施	
		安全用电					
		布线规范、合理					
		工具摆放整齐					
		工具及仪器仪表使用规范、摆放整齐					
		任务完成后,进行场地整理,保持场地清洁、有序					
2	实训态度	不迟到、早退、旷课		10%	现场考评	六步	
		实训过程认真负责					
		组内主动沟通、协作,小组间互助					
3	系统方案制定	工作流程正确合理		10%	现场考评	计划决策	
		方案合理					
		选用指令是否合理					
		电路图正确					
	编程能力	独立完成程序		10%	现场考评	决策	
		程序简单、可靠					

（续表）

4	操作能力	正确输入程序并进行程序调试	20%	现场考评	实施
		根据电路图正确接线			
		根据系统功能进行正确操作演示			
5	工艺	接线美观	10%	现场考评	实施
		线路工作可靠			
6	实践效果	系统工作可靠	10%	现场考评	检查
		满足工作要求			
		创新			
		按规定的时间完成项目			
7	汇报总结	工作总结，PPT汇报	5%	现场考评	评估
		填写自我检查表及反馈表			
8	技术文件制作整理	技术文件制作整理能力	5%	现场考评	评估
		合计	100%		

表7-3 各部分考核占课程考核的比例

考核项目	理论考核	技能考核	职业素质考核	合 计
比 例	30%	50%	20%	100%
分 值	30	50	20	100
实际得分				

任务 8 广告灯控制

实训任务

广告灯的左移右移,八个发光二极管 L1~L8 分别接在单片机的 P0.0~P0.7 接口上,输出"0"时,发光二极管亮,开始时 P0.0→P0.1→P0.2→P0.3→…→P0.7→P0.6→…→P0.0 亮,重复循环。

实训设备

1. 设备

PC 机(安装 wave 编程软件、Keil C51 软件)、单片机实验板。

2. 工具及材料

工作对象:电工电子工具、电子元器件和辅助材料、仿真器、编程器。

工作工具:单片机控制电路原理图、实训指导书、项目任务单、工作记录单、项目检查单、各种电工仪表、常用电工工具和拆装工具、量具、相关电子手册。

硬件设计

主控模块采用 ATMEL 公司生产的 AT89S52 单片机。广告灯模块与单片机的接口电路如图 8-1 所示。

软件设计

一、设计思想

我们可以运用输出端口指令 MOV P0,A 或 MOV P0,♯DATA,只要给累加器值或常数值,然后执行上述的指令,即可达到输出控制的动作。每次送出的数据是不同,具体的数据见表 8-1。

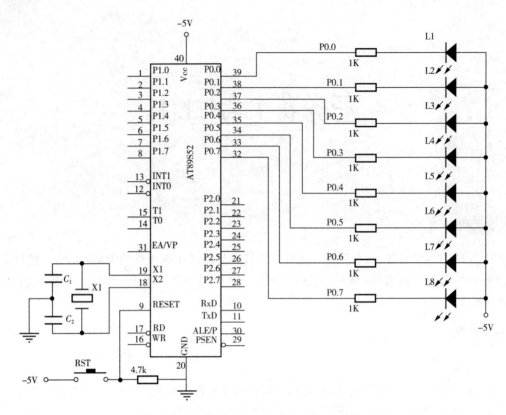

图 8-1　广告灯模块与单片机的接口电路图

表 8-1　广告灯的左移右移数据表

P0.7	P0.6	P0.5	P0.4	P0.3	P0.2	P0.1	P0.0	说明
L8	L7	L6	L5	L4	L3	L2	L1	
1	1	1	1	1	1	1	0	L1 亮
1	1	1	1	1	1	0	1	L2 亮
1	1	1	1	1	0	1	1	L3 亮
1	1	1	1	0	1	1	1	L4 亮
1	1	1	0	1	1	1	1	L5 亮
1	1	0	1	1	1	1	1	L6 亮
1	0	1	1	1	1	1	1	L7 亮
0	1	1	1	1	1	1	1	L8 亮

二、程序流程图

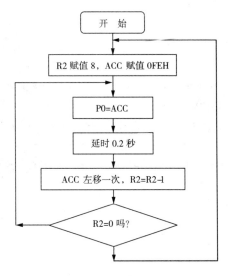

图 8-2 程序流程图

三、汇编语言源程序

```
;****************************************************************************
;项目名称:广告灯左移右移控制
;功能:循环点亮广告灯 P0.0→P0.1→P0.2→P0.3→…→P0.7→P0.6→…→P0.0
;****************************************************************************
        ORG    0000H
START: MOV    R2,#8
        MOV    A,#0FEH
        SETB   C
LOOP:  MOV    P0,A
        LCALL  DELAY
        RLC    A
        DJNZ   R2,LOOP
        MOV    R2,#8
LOOP1: MOV    P0,A
        LCALL  DELAY
        RRC    A
        DJNZ   R2,LOOP1
        LJMP   START
;延时子程序
DELAY: MOV    R5,#20
D1:    MOV    R6,#20
D2:    MOV    R7,#248
        DJNZ   R7,$
        DJNZ   R6,D2
```

```
        DJNZ  R5,D1
        RET
        END
```

四、C语言源程序

```c
//*********************************************
//项目名称:广告灯左移右移控制
//功能:循环点亮广告灯 P0.0→P0.1→P0.2→P0.3→…→P0.7→P0.6→…→P0.0
//*********************************************
#include<AT89X51.H>
unsigned char i;
unsigned char temp;
unsigned char a,b;
void delay(void)
{
    unsigned char m,n,s;
    for(m=20;m>0;m--)
      for(n=20;n>0;n--)
         for(s=248;s>0;s--);
}
void main(void)      //主函数
{
    while(1)
    {
        temp=0xfe;
        P0=temp;
        delay();
        for(i=1;i<8;i++)
        {
            a=temp<<i;
            b=temp>>(8-i);
            P0=a|b;
            delay();
        }
        for(i=1;i<8;i++)
        {
            a=temp>>i;
            b=temp<<(8-i);
            P0=a|b;
            delay();
        }
```

实训考核

本课程改革传统的闭卷或开卷考核,而采用过程考核为主的多元化考核方式,考核分为理论考核、职业道德考核和技能考核三部分,各部分所占比例见表8-2、表8-3、表8-4。

表8-2 理论考核和职业素质考核形式及所占比例

序号	名称		比例	得分
一	理论考核	过程作业文件 个人自评	10%	
		过程作业文件 组内互评	20%	
		过程作业文件 小组互评	30%	
		过程作业文件 老师评定	20%	
		课堂提问、解答	10%	
		项目汇报	10%	
		小计	100%	
二	职业素质	职业道德,工作作风	40%	
		小组沟通协作能力	40%	
		创新能力	20%	
		小计	100%	

表8-3 技能考核内容及比例

姓名		班级		小组		总得分	
序号	考核项目	考核内容及要求		配分	评分标准	考核环节	得分
1	①安全文明生产 ②安全操作规范	着装规范		20%	现场考评	实施	
		安全用电					
		布线规范、合理					
		工具摆放整齐					
		工具及仪器仪表使用规范、摆放整齐					
		任务完成后,进行场地整理,保持场地清洁、有序					
2	实训态度	不迟到、早退、旷课		10%	现场考评	六步	
		实训过程认真负责					
		组内主动沟通、协作,小组间互助					

（续表）

序号	考核项目	考核内容	比例	考核方式	能力目标
3	系统方案制定	工作流程正确合理	10%	现场考评	计划决策
		方案合理			
		选用指令是否合理			
		电路图正确			
	编程能力	独立完成程序	10%	现场考评	决策
		程序简单、可靠			
4	操作能力	正确输入程序并进行程序调试	20%	现场考评	实施
		根据电路图正确接线			
		根据系统功能进行正确操作演示			
5	工艺	接线美观	10%	现场考评	实施
		线路工作可靠			
6	实践效果	系统工作可靠	10%	现场考评	检查
		满足工作要求			
		创新			
		按规定的时间完成项目			
7	汇报总结	工作总结，PPT汇报	5%	现场考评	评估
		填写自我检查表及反馈表			
8	技术文件制作整理	技术文件制作整理能力	5%	现场考评	评估
	合计		100%		

表8-4 各部分考核占课程考核的比例

考核项目	理论考核	技能考核	职业素质考核	合 计
比 例	30%	50%	20%	100%
分 值	30	50	20	100
实际得分				

任务 9　流水灯控制

实训任务

利用取表的方法,使端口 P0 做流水灯的变化:左移 2 次,右移 2 次,闪烁 2 次(延时的时间 0.2 秒)。

实训设备

1. 设备

PC 机(安装 wave 编程软件、Keil C51 软件)、单片机实验板。

2. 工具及材料

工作对象:电工电子工具、电子元器件和辅助材料、仿真器、编程器。

工作工具:单片机控制电路原理图、实训指导书、项目任务单、工作记录单、项目检查单、各种电工仪表、常用电工工具和拆装工具、量具、相关电子手册。

硬件设计

主控模块采用 ATMEL 公司生产的 AT89S52 单片机。流水灯模块与单片机的接口电路如图 9-1 所示。

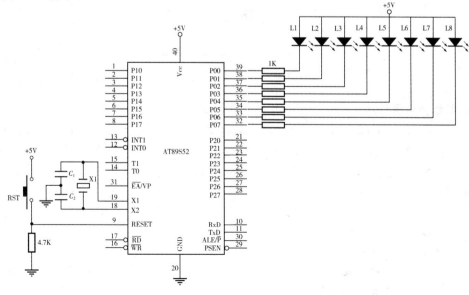

图 9-1　流水灯模块与单片机的接口电路

软件设计

一、设计思想

在用表格进行程序设计的时候，要用以下的指令来完成。

(1)利用 MOV DPTR,♯DATA16 的指令来使数据指针寄存器指到表的开头。

(2)利用 MOVC A,@A+DPTR 的指令,根据累加器的值再加上 DPTR 的值,就可以使程序计数器 PC 指到表格内所要取出的数据。

因此,只要把控制码建成一个表,而利用 MOVC A,@A+DPTR 做取码的操作,就可方便地处理一些复杂的控制动作,取表过程如图 9-2 所示。

二、程序流程图

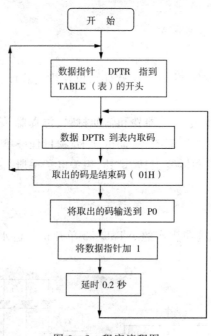

图 9-2 程序流程图

三、汇编语言源程序

```
;******************************************************************
;项目名称:流水灯控制
;功能:P0 端口的流水灯如下变化:左移 2 次,右移 2 次,闪烁 2 次
;******************************************************************
        ORG     0000H
START:  MOV     DPTR,♯TABLE
LOOP:   CLR     A
        MOVC    A,@A+DPTR
        CJNE    A,♯01H,LOOP1
        JMP     START
```

```
LOOP1: MOV    P0,A
       MOV    R3,#20
       LCALL  DELAY
       INC    DPTR
       JMP    LOOP
;延时子程序
DELAY: MOV    R4,#20
D1:    MOV    R5,#248
       DJNZ   R5,$
       DJNZ   R4,D1
       DJNZ   R3,DELAY
       RET
TABLE: DB 0FEH,0FDH,0FBH,0F7H
       DB 0EFH,0DFH,0BFH,07FH
       DB 0FEH,0FDH,0FBH,0F7H
       DB 0EFH,0DFH,0BFH,07FH
       DB 07FH,0BFH,0DFH,0EFH
       DB 0F7H,0FBH,0FDH,0FEH
       DB 07FH,0BFH,0DFH,0EFH
       DB 0F7H,0FBH,0FDH,0FEH
       DB 00H,0FFH,00H,0FFH
       DB 01H
       END
```

四、C语言源程序

```c
//****************************************************************************
//项目名称:流水灯控制
//功能:P0端口的流水灯如下变化:左移2次,右移2次,闪烁2次
//****************************************************************************
#include<AT89X51.H>
unsigned char code table[] = {0xfe,0xfd,0xfb,0xf7,0xef,0xdf,0xbf,0x7f,0xfe,0xfd,0xfb,
0xf7,0xef,0xdf,0xbf,0x7f,0x7f,0xbf,0xdf,0xef,0xf7,0xfb,0xfd,0xfe,0x7f,0xbf,0xdf,0xef,0xf7,
0xfb,0xfd,0xfe,0x00,0xff,0x00,0xff,0x01};
unsigned char i;
void delay(void)
{
    unsigned char m,n,s;
    for(m=20;m>0;m--)
    for(n=20;n>0;n--)
    for(s=248;s>0;s--);
}
//主函数
void main(void)
```

```
    {
        while(1)
        {
            if(table[i]! = 0x01)
            {
                P0 = table[i];
                i + + ;
                delay( );
            }
            else   {i=0;}
        }
    }
```

实训考核

本课程改革传统的闭卷或开卷考核,而采用过程考核为主的多元化考核方式,考核分为理论考核、职业道德考核和技能考核三部分,各部分所占比例见表9-1、表9-2、表9-3。

表9-1 理论考核和职业素质考核形式及所占比例

序 号	名 称		比 例	得 分
一	理论考核	过程作业文件	个人自评	10%	
			组内互评	20%	
			小组互评	30%	
			老师评定	20%	
		课堂提问、解答	10%		
		项目汇报	10%		
		小计	100%		
二	职业素质	职业道德,工作作风	40%		
		小组沟通协作能力	40%		
		创新能力	20%		
		小计	100%		

表9-2 技能考核内容及比例

姓名		班级		小组		总得分	
序号	考核项目	考核内容及要求		配分	评分标准	考核环节	得分
1	①安全文明生产 ②安全操作规范	着装规范		20%	现场考评	实施	
		安全用电					
		布线规范、合理					
		工具摆放整齐					
		工具及仪器仪表使用规范、摆放整齐					
		任务完成后,进行场地整理,保持场地清洁、有序					
2	实训态度	不迟到、早退、旷课		10%	现场考评	六步	
		实训过程认真负责					
		组内主动沟通、协作,小组间互助					
3	系统方案制定	工作流程正确合理		10%	现场考评	计划决策	
		方案合理					
		选用指令是否合理					
		电路图正确					
	编程能力	独立完成程序		10%	现场考评	决策	
		程序简单、可靠					
4	操作能力	正确输入程序并进行程序调试		20%	现场考评	实施	
		根据电路图正确接线					
		根据系统功能进行正确操作演示					
5	工艺	接线美观		10%	现场考评	实施	
		线路工作可靠					
6	实践效果	系统工作可靠		10%	现场考评	检查	
		满足工作要求					
		创新					
		按规定的时间完成项目					
7	汇报总结	工作总结,PPT汇报		5%	现场考评	评估	
		填写自我检查表及反馈表					
8	技术文件制作整理	技术文件制作整理能力		5%	现场考评	评估	
		合计		100%			

表9-3 各部分考核占课程考核的比例

考核项目	理论考核	技能考核	职业素质考核	合　计
比　例	30%	50%	20%	100%
分　值	30	50	20	100
实际得分				

任务 10 数码管显示技术

实训任务

利用 AT89S52 单片机的 P0 端口的 P0.0～P0.7 连接到一个共阳数码管的 a～dp 的字段上,数码管的公共端使能 P2.0 接到＋5V。在数码管上循环显示 0～9 数字,时间间隔 0.2 秒。

实训设备

1. 设备

PC 机(安装 wave 编程软件、Keil C51 软件)、单片机实验板。

2. 工具及材料

工作对象:电工电子工具、电子元器件和辅助材料、仿真器、编程器。

工作工具:单片机控制电路原理图、实训指导书、项目任务单、工作记录单、项目检查单、各种电工仪表、常用电工工具和拆装工具、量具、相关电子手册。

硬件设计

主控模块采用 ATMEL 公司生产的 AT89S52 单片机。选用 8 段共阳极数码管,数码管的 a、b、c、d、e、f、g、dp 段分别与单片机 P0 口的 P0.0、P0.1、P0.2、P0.3、P0.4、P0.5、P0.6、P0.7 相连,用来控制显示数字的形状。数码管的公共使能端 COM 连接三极管 C8550 的集电极,三极管 C8550 主要用于信号的放大,以驱动数码管工作。三极管 8550 的基极通过限流电阻接到单片机 P2 口的 P2.0,通过控制三极管 C8550 的基极电平来打开或关闭数码管的显示,起到"使能"作用。三极管 C8550 的集电极接＋5V 电源。数码管显示模块与单片机的接口电路如图 10-1 所示。

软件设计

一、算法设计

对于数码管而言,要想显示数字或字母,首先应该选中该数码管,然后相应字段被点亮。例如:显示一个"3"字,那么应当是 a 亮、b 亮、c 亮、d 亮、e 不亮、f 不亮、g 亮、dp 不亮。对于共阳极数码管,对应到单片机的 P0 口,P0.7 为高电平"0",P0.6 为低电平"1",P0.5 为高电平"0",P0.4 为高电平"1",P0.3 为低电平"0",P0.2 为低电平"1",P0.1 为低电平"0",P0.0 为低电平"0",即当把 1011 0100(0Xb0)送给 P0 口时,可以显示"3"字。

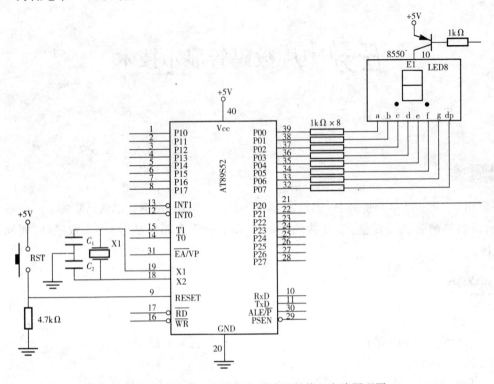

图 10-1 数码管显示模块与单片机的接口电路原理图

其他数字的显示依此类推,那么可以列出数码管显示数字的字形码表见表 10-1。

表 10-1 数码管显示数字的字形码表

显示字符	共阳极								字形码
	D7	D6	D5	D4	D3	D2	D1	D0	
	dp	g	f	e	d	c	b	a	
0	1	1	0	0	0	0	0	0	C0H
1	1	1	1	1	1	0	0	1	F9H
2	1	0	1	0	0	1	0	0	A4H
3	1	0	1	1	0	0	0	0	B0H
4	1	0	0	1	1	0	0	1	99H
5	1	0	0	1	0	0	1	0	92H
6	1	0	0	0	0	0	1	0	82H
7	1	1	1	1	1	0	0	0	F8H
8	1	0	0	0	0	0	0	0	80H
9	1	0	0	1	0	0	0	0	90H

在表 10-1 中，由于数码管显示的数字"0～9"的字形码"0xc0,0xf9,0xa4,0xb0,0x99,0x92,0x82,0xf8,0x80,0x90"没有规律可循，可以采用数组的方式来完成我们所需的要求。

有时候我们需要把一些数据表存放在数组中，通常这些数据是不用在程序中改变数值的，这个时候就要把这些数据在程序编写时就赋给数组变量。因为 51 芯片的片内 RAM 很有限，通常会把 RAM 分给参与运算的变量或数组，而那些程序中不变数据则应存放在片内的 CODE 存储区，以节省宝贵的 RAM。

在程序设计中可以设计一个变量表示数码管字形码数组索引值，每隔一定时间在"0～9"之间变化，然后按照这个数组索引值去查找相应的数码管字形码数组值，并把查到的数据送到 P0 口进行显示。

建立数组变量的程序代码如下所示：

unsigned char codeledcode[ ] = {0xc0,0xf9,0xa4,0xb0,0x99,0x92,0x82,0xf8,0x80,0x90}。

二、程序设计

1. 主函数设计

主函数主要完成硬件初始化、函数调用等功能。

(1) 初始化

设置数码管使能信号 P2.0 有效，P2.0＝0，即 P2＝0xfe；将数码管字形码数组索引值 dispcount 赋初值 0。

(2) 字符显示

根据数码管字形码数组索引值 dispcount，取得相应数组值 ledcode[dispcount]，把数组值 ledcode[dispcount]送给 P0，即可在数码管上显示相应的数字，并调用 0.2s 延时函数（为了使数码管能够稳定显示相应的数字）；最后，数码管字形码数组索引值自加 1，为下一个数字的显示做准备。

(3) 显示结束判断

对数码管字形码数组索引值 dispcount 进行判断，看它是否小于 10：

若 dispcount 值小于 10，表示本轮数字未显示完成，则程序跳转到"字符显示"处执行，数码管显示下一个数字。

若 dispcount 值不小于 10，表示本轮数字显示完成，程序跳转至主程序开始，重新数码管字形码数组索引值 dispcount 值赋初值 0，进行下一轮显示，从而实现数字 0～9 的循环显示。

主函数设计流程图如图 10-2 所示。

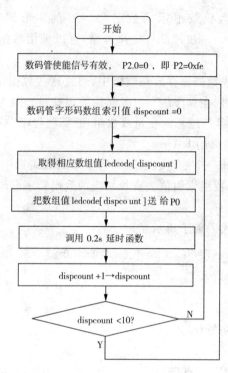

图 10-2 主函数设计流程图

三、汇编语言源程序

;**
;项目名称:数码管显示技术
;功能:在数码管上循环显示数字 0~9
;**
//主程序

```
        ORG    0000H              ;程序入口地址
        MOV    P2,#0FEH           ;P2.0=0,数码管使能信号 P2.0 有效
START:  MOV    R1,#00H            ;将数码管数据表索引寄存器 R1 赋初值 00H
NEXT:   MOV    A,R1               ;把字形码的变址地址送给累加器 A
        MOV    DPTR,#TABLE        ;取得表首地址
        MOVC   A,@A+DPTR          ;查表,取得显示的字型码送给累加器 A
        MOV    P0,A               ;把字型码送到 P0 进行显示
        LCALL  DELAY200ms         ;调用延时子程序 DELAY200ms
        INC    R1                 ;下一个字形码的变址寄存器
        CJNE   R1,#10,NEXT        ;判断 0~9 是否显示完毕
        LJMP   START              ;进入下一轮循环显示 0~9
;延时 0.2s 子程序 DELAY200ms
DELAY200ms: MOV R5,#20            ;外层循环结构开始,设置外层循环变量初始值
LOOP1:      MOV R6,#20            ;中层循环结构开始,设置中层循环变量初始值
LOOP2:      MOV R7,#230           ;内层循环结构开始,设置内层循环变量初始值
            DJNZ R7,$             ;内层循环结束判断条件,直到 R7=0 为止
```

```
        DJNZ R6,    LOOP2        ;中层循环结束判断条件,直到 R6＝0 为止
        DJNZ R5,    LOOP1        ;外层循环结束判断条件,直到 R5＝0 为止
        RET                      ;子程序返回
;LED 字形码表
TABLE:DB0C0H,0F9H,0A4H,0B0H,99H,92H,82H,0F8H,80H,90H
END
```

四、C 语言源程序

```
//**********************************************************************
//项目名称:数码管显示技术
//功能:在数码管上循环显示数字 0～9
//**********************************************************************
#include<AT89X51.H>                  //包含"51 寄存器定义"头文件
//变量定义:ledcode[ ]—存储数码管字形码
//程序中不变数据存放在片内的 CODE 存储区,以节省宝贵的 RAM
unsigned char codeledcode[ ] = {0xc0,0xf9,0xa4,0xb0,0x99,0x92,0x82,0xf8,0x80,0x90,0x88,
0x83,0xc6,0xa1,0x86,0x8e};
unsigned char dispcount;              //数码管字形码数组索引值 dispcount
//函数声明:延时 0.2s 函数 delay02s()
void delay02s(void);
//主函数
void main(void)                       //主函数入口
{
    P2 = 0xfe;                        //P2.0＝0,数码管使能信号 P2.0 有效
    while(1)                          //进入下一轮数码管循环显示 0～9
    {
        //判断每轮数码管循环显示 0～9 是否显示完毕
        for(dispcount = 0;dispcount<10;dispcount + + )
        {
            P0 = ledcode[dispcount];  //数码管显示值 ledcode[dispcount]送给 P0
            delay02s();               //调用延时子程序 delay02s()
        }
    }
}
//延时 0.2s 函数 delay02s()
void delay02s(void)
{
    unsigned char i,j,k;              //定义无符号字符型变量 i,j,k
    for(i = 20;i>0;i - - )            //外层循环
        for(j = 20;j>0;j - - )        //中层循环
            for(k = 230;k>0;k - - );  //内层循环
}
```

实训考核

本课程改革传统的闭卷或开卷考核,而采用过程考核为主的多元化考核方式,考核分为理论考核、职业道德考核和技能考核三部分,各部分所占比例见表 10-2、表 10-3、表10-4。

表 10-2 理论考核和职业素质考核形式及所占比例

序 号	名 称		比 例	得 分
一	理论考核	过程作业文件	个人自评	10%	
			组内互评	20%	
			小组互评	30%	
			老师评定	20%	
		课堂提问、解答		10%	
		项目汇报		10%	
		小计		100%	
二	职业素质	职业道德,工作作风		40%	
		小组沟通协作能力		40%	
		创新能力		20%	
		小计		100%	

表 10-3 技能考核内容及比例

姓名		班级		小组		总得分	
序号	考核项目	考核内容及要求		配分	评分标准	考核环节	得分
1	①安全文明生产 ②安全操作规范	着装规范		20%	现场考评	实施	
		安全用电					
		布线规范、合理					
		工具摆放整齐					
		工具及仪器仪表使用规范、摆放整齐					
		任务完成后,进行场地整理,保持场地清洁、有序					
2	实训态度	不迟到、早退、旷课		10%	现场考评	六步	
		实训过程认真负责					
		组内主动沟通、协作,小组间互助					

(续表)

3	系统方案制定	工作流程正确合理	10%	现场考评	计划决策
		方案合理			
		选用指令是否合理			
		电路图正确			
	编程能力	独立完成程序	10%	现场考评	决策
		程序简单、可靠			
4	操作能力	正确输入程序并进行程序调试	20%	现场考评	实施
		根据电路图正确接线			
		根据系统功能进行正确操作演示			
5	工艺	接线美观	10%	现场考评	实施
		线路工作可靠			
6	实践效果	系统工作可靠	10%	现场考评	检查
		满足工作要求			
		创新			
		按规定的时间完成项目			
7	汇报总结	工作总结，PPT 汇报	5%	现场考评	评估
		填写自我检查表及反馈表			
8	技术文件制作整理	技术文件制作整理能力	5%	现场考评	评估
	合计		100%		

表 10-4 各部分考核占课程考核的比例

考核项目	理论考核	技能考核	职业素质考核	合　计
比　例	30%	50%	20%	100%
分　值	30	50	20	100
实际得分				

任务 11 动态数码显示技术

实训任务

单片机 P0 端口接动态数码管的字形码笔段,P2 端口接动态数码管的数位选择端,P1.0 接一个开关,当开关接高电平时,显示"12345678"字样;当开关接低电平时,显示"87654321"字样。

实训设备

1. 设备
PC 机(安装 wave 编程软件、Keil C51 软件)、单片机实验板。
2. 工具及材料
工作对象:电工电子工具、电子元器件和辅助材料、仿真器、编程器。
工作工具:单片机控制电路原理图、实训指导书、项目任务单、工作记录单、项目检查单、各种电工仪表、常用电工工具和拆装工具、量具、相关电子手册。

硬件设计

主控模块采用 ATMEL 公司生产的 AT89S52 单片机。选用 8 段共阳极数码管,数码管的 a、b、c、d、e、f、g、dp 段分别与单片机 P0 口的 P0.0、P0.1、P0.2、P0.3、P0.4、P0.5、P0.6、P0.7 相连,用来控制显示数字的形状。数码管的公共使能端 COM 连接三极管 C8550 的集电极,三极管 C8550 主要用于信号的放大,以驱动数码管工作。三极管 8550 的基极通过限流电阻接到单片机 P2 口的 P2.0,通过控制三极管 C8550 的基极电平来打开或关闭数码管的显示,起到"使能"作用。三极管 C8550 的集电极接+5V 电源。动态数码显示模块与单片机的接口电路如图 11-1 所示。

软件设计

一、数码管动态扫描方法

(1)动态接口采用各数码管循环轮流显示的方法,当循环显示频率较高时,利用人眼的暂留特性,看不出闪烁显示现象,这种显示需要一个接口完成字形码的输出(字形选择),另一接口完成各数码管的轮流点亮(数位选择)。

(2)在进行数码显示的时候,要对显示单元开辟 8 个显示缓冲区,每个显示缓冲区装有显示的不同数据即可。

(3)对于显示的字形码数据采用查表方法来完成。

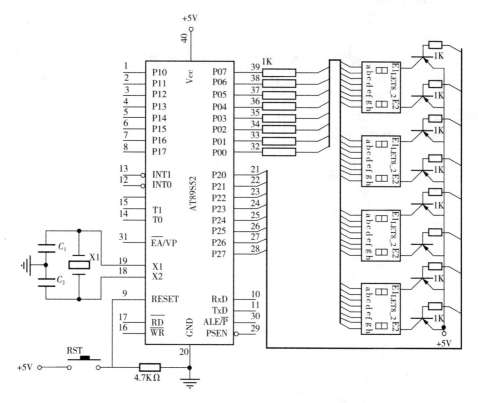

图 11-1 动态数码显示模块与单片机的接口电路

二、程序流程图

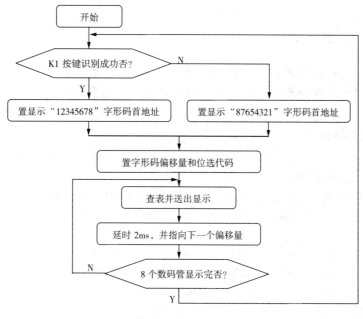

图 11-2 程序流程图

三、汇编语言源程序

```
;*******************************************************************
;项目名称:动态数码显示
;功能:当开关接高电平时,显示"12345678";当开关接低电平时,显示"87654321"。
;*******************************************************************
        ORG     0000H
START:  JB      P1.0,DIR1
        MOV     DPTR,#TABLE1
        SJMP    DIR
DIR1:   MOV     DPTR,#TABLE2
DIR:    MOV     R0,#00H
        MOV     R1,#0FEH
NEXT:   MOV     A,R0
        MOVC    A,@A+DPTR
        MOV     P0,A
        MOV     A,R1
        MOV     P2,A
        LCALL   DAY
        INC     R0
        RL      A
        MOV     R1,A
        CJNE    R1,#0FEH,NEXT
        SJMP    START
DAY:    MOV     R6,#4
D1:     MOV     R7,#248
        DJNZ    R7,$
        DJNZ    R6,D1
        RET
TABLE1:DB 0F9H,0A4H,0B0H,99H,92H,82H,0F8H,80H    ;显示 1--8
TABLE2:DB 80H,0F8H,82H,92H,99H,0B0H,0A4H,0F9H    ;显示 8--1
END
```

四、C 语言源程序

```c
//*******************************************************************
//项目名称:动态数码显示
//功能:当开关接高电平时,显示"12345678";当开关接低电平时,显示"87654321"。
//*******************************************************************
#include<reg51.H>
unsigned char code table1[]={0xf9,0xa4,0xb0,0x99,0x92,0x82,0xf8,0x80};
unsigned char code table2[]={0x80,0xf8,0x82,0x92,0x99,0xb0,0xa4,0xf9};
unsigned char i;
unsigned char t1,t2;
unsigned char temp;
```

```
sbit K1 = P1^0;
//主函数
void main(void)
{
    while(1)
    {
        temp = 0xfe;
        for(i = 0;i<8;i++)
        {
            if(K1 = = 1)   {P0 = table1[i];}
            else           {P0 = table2[i];}
            P2 = temp;
            temp = (temp<<1)|0x01;
            for(t1 = 0;t1<4;t1++)
            {  for(t2 = 0;t2<200;t2++)
                {}
            }
        }
    }
}
```

实训考核

本课程改革传统的闭卷或开卷考核,而采用过程考核为主的多元化考核方式,考核分为理论考核、职业道德考核和技能考核三部分,各部分所占比例见表 11 - 1、表 11 - 2、表11 - 3。

表 11 - 1 理论考核和职业素质考核形式及所占比例

序号	名称		比例	得分
一	理论考核	过程作业文件	个人自评	10%	
			组内互评	20%	
			小组互评	30%	
			老师评定	20%	
		课堂提问、解答	10%	
		项目汇报	10%	
		小计	100%	
二	职业素质	职业道德,工作作风	40%	
		小组沟通协作能力	40%	
		创新能力	20%	
		小计	100%	

表 11-2 技能考核内容及比例

姓名		班 级		小组		总得分	
序号	考核项目	考核内容及要求	配分	评分标准	考核环节	得分	
1	①安全文明生产　②安全操作规范	着装规范	20%	现场考评	实施		
		安全用电					
		布线规范、合理					
		工具摆放整齐					
		工具及仪器仪表使用规范、摆放整齐					
		任务完成后,进行场地整理,保持场地清洁、有序					
2	实训态度	不迟到、早退、旷课	10%	现场考评	六步		
		实训过程认真负责					
		组内主动沟通、协作,小组间互助					
3	系统方案制定	工作流程正确合理	10%	现场考评	计划决策		
		方案合理					
		选用指令是否合理					
		电路图正确					
	编程能力	独立完成程序	10%	现场考评	决策		
		程序简单、可靠					
4	操作能力	正确输入程序并进行程序调试	20%	现场考评	实施		
		根据电路图正确接线					
		根据系统功能进行正确操作演示					
5	工艺	接线美观	10%	现场考评	实施		
		线路工作可靠					
6	实践效果	系统工作可靠	10%	现场考评	检查		
		满足工作要求					
		创新					
		按规定的时间完成项目					
7	汇报总结	工作总结,PPT 汇报	5%	现场考评	评估		
		填写自我检查表及反馈表					
8	技术文件制作整理	技术文件制作整理能力	5%	现场考评	评估		
		合计	100%				

表 11-3 各部分考核占课程考核的比例

考核项目	理论考核	技能考核	职业素质考核	合　计
比　例	30%	50%	20%	100%
分　值	30	50	20	100
实际得分				

任务 12 独立按键识别技术

实训任务

每按下一次开关 K1，计数值加 1，通过单片机的 P0 端口的 P0.0 到 P0.7 显示其二进制计数值。

实训设备

1. 设备

PC 机（安装 wave 编程软件、Keil C51 软件）、单片机实验板。

2. 工具及材料

工作对象：电工电子工具、电子元器件和辅助材料、仿真器、编程器。

工作工具：单片机控制电路原理图、实训指导书、项目任务单、工作记录单、项目检查单、各种电工仪表、常用电工工具和拆装工具、量具、相关电子手册。

硬件设计

主控模块采用 ATMEL 公司生产的 AT89S52 单片机。独立按键模块与单片机的接口电路如图 12-1 所示。

软件设计

一、独立按键

一个按键从没有按下到按下以及释放是一个完整的过程。当我们按下一个按键时，总希望某个命令只执行一次，而在按键按下的过程中，不要有干扰。因为，在按下的过程中，一旦有干扰过来，可能造成误触发过程。因此在按键按下的时候，要把干扰信号以及按键的机械接触等干扰信号给滤除掉。可以采用电容来滤除掉这些干扰信号，但会增加硬件成本及硬件电路的体积。因此可以采用软件滤波的方法去除这些干扰信号，一个按键按下的时刻存在着一定的干扰信号，按下之后就基本上进入了稳定的状态。具体的一个按键从按下到释放的全过程的信号图如图 12-2 所示。

任务 12 独立按键识别技术

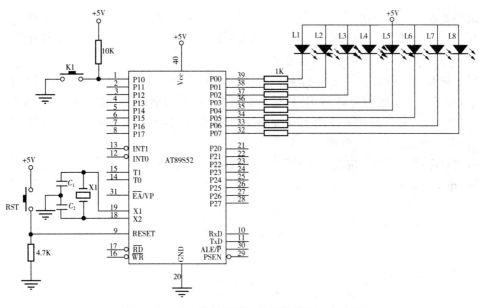

图 12-1 独立按键模块与单片机的接口电路图

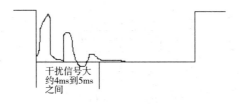

图 12-2 按键过程示意图

从图中可以看出,在程序设计时,从按键被识别按下之后,延时 5ms 以上,从而避开了干扰信号区域。我们再来检测一次,看按键是否真的已经按下。若已经按下,这时肯定输出为低电平,若这时检测到的是高电平,证明刚才是由于干扰信号引起的误触发,CPU 就认为是误触发信号而舍弃这次的按键识别过程,从而提高了系统的可靠性。

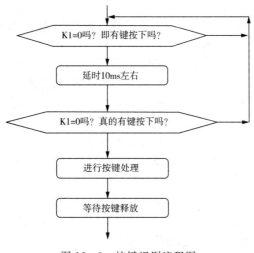

图 12-3 按键识别流程图

· 81 ·

由于要求每按下一次,命令被执行一次,直到下一次再按下的时候,再执行一次命令。因此从按键被识别出来之后,就可以执行这次的命令,所以要有一个等待按键释放的过程,显然释放的过程,就是使其恢复成高电平状态。

对于按键识别的指令,依然选择如下指令 JB BIT,REL 指令是用来检测 BIT 是否为高电平,若 BIT=1,则程序转向 REL 处执行程序,否则就继续向下执行程序。或者是 JNB BIT,REL 指令是用来检测 BIT 是否为低电平,若 BIT=0,则程序转向 REL 处执行程序,否则就继续向下执行程序。

但对程序设计过程中按键识别过程的框图如图 12-3 所示。

二、程序流程图

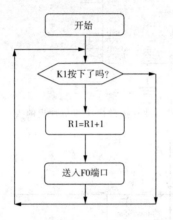

图 12-4 程序流程图

三、汇编语言源程序

```
;*********************************************************************
;项目名称:独立按键识别技术
;功能:每按下一次开关 K1,计数值加 1,通过单片机的 P0 口显示其二进制计数值。
;*********************************************************************
        ORG     0000H
START:  MOV     R0,#00H         ;初始化 R0 为 0,表示从 0 开始计数
        MOV     A,R0            ;
        CPL     A               ;取反指令
        MOV     P0,A            ;送出 P0 端口由发光二极管显示
REL:    JB      P1.0,REL        ;判断 K1 是否按下
        LCALL   DELAY10MS       ;若按下,则延时 10ms 左右
        JB      P1.0,REL        ;再判断 K1 是否真得按下
        INC     R0              ;若真得按下,则进行按键处理,使
        MOV     A,R0            ;计数内容加 1,并送出 P0 端口由
        CPL     A               ;发光二极管显示
        MOV     P0,A            ;
        JNB     P1.0,$          ;等待 K1 释放
        SJMP    REL             ;继续对 K1 按键扫描
```

;延时10ms子程序
```
DELAY10MS:   MOV  R6,#20
L1:          MOV  R7,#230
             DJNZ R7,$
             DJNZ R6,L1
             RET
             END
```

四、C语言源程序

```c
// ****************************************************************************
//项目名称:独立按键识别技术
//功能:每按下一次开关K1,计数值加1,通过单片机的P0口显示其二进制计数值。
// ****************************************************************************
#include<AT89X51.H>
unsigned char count;
//延时函数
void delay10ms(void)
{
    unsigned char i,j;
    for(i=20;i>0;i--)
        for(j=248;j>0;j--);
}
//主函数
void main(void)
{
    while(1)
    {
        if(P1_0==0)
        {
            delay10ms();
            if(P1_0==0)
            {
                count++;
                P0=~count;
                while(P1_0==0);
            }
        }
    }
}
```

实训考核

本课程改革传统的闭卷或开卷考核,而采用过程考核为主的多元化考核方式,考核分为理论考核、职业道德考核和技能考核三部分,各部分所占比例见表12-1、表12-2、表12-3。

表 12-1 理论考核和职业素质考核形式及所占比例

序号	名称		比例	得分
一	理论考核	个人自评	10%	
		组内互评	20%	
		过程作业文件 小组互评	30%	
		老师评定	20%	
		课堂提问、解答	10%	
		项目汇报	10%	
		小计	100%	
二	职业素质	职业道德,工作作风	40%	
		小组沟通协作能力	40%	
		创新能力	20%	
		小计	100%	

表 12-2 技能考核内容及比例

姓名		班级		小组		总得分	
序号	考核项目	考核内容及要求		配分	评分标准	考核环节	得分
1	①安全文明生产 ②安全操作规范	着装规范		20%	现场考评	实施	
		安全用电					
		布线规范、合理					
		工具摆放整齐					
		工具及仪器仪表使用规范、摆放整齐					
		任务完成后,进行场地整理,保持场地清洁、有序					
2	实训态度	不迟到、不早退、不旷课		10%	现场考评	六步	
		实训过程认真负责					
		组内主动沟通、协作,小组间互助					
3	系统方案制定	工作流程正确合理		10%	现场考评	计划决策	
		方案合理					
		选用指令是否合理					
		电路图正确					
	编程能力	独立完成程序		10%	现场考评	决策	
		程序简单、可靠					

(续表)

4	操作能力	正确输入程序并进行程序调试	20%	现场考评	实施	
		根据电路图正确接线				
		根据系统功能进行正确操作演示				
5	工艺	接线美观	10%	现场考评	实施	
		线路工作可靠				
6	实践效果	系统工作可靠	10%	现场考评	检查	
		满足工作要求				
		创新				
		按规定的时间完成项目				
7	汇报总结	工作总结,PPT汇报	5%	现场考评	评估	
		填写自我检查表及反馈表				
8	技术文件制作整理	技术文件制作整理能力	5%	现场考评	评估	
	合计		100%			

表 12-3 各部分考核占课程考核的比例

考核项目	理论考核	技能考核	职业素质考核	合 计
比 例	30%	50%	20%	100%
分 值	30	50	20	100
实际得分				

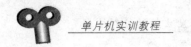

任务 13 一键多功能按键识别技术

实训任务

开关 K1 接在 P1.0 管脚上，在单片机的 P0 端口接有四个发光二极管。系统上电的时候，接在 P0.0 管脚上的发光二极管 L1 闪烁；当每一次按下开关 K1 的时候，接在 P0.1 管脚上的发光二极管 L2 闪烁；再按下开关 K1 的时候，接在 P0.2 管脚上的发光二极管 L3 在闪烁，再按下开关 K1 的时候，接在 P0.3 管脚上的发光二极管 L4 闪烁；再按下开关 K1 的时候，又轮到 L1 闪烁，如此轮流循环。

实训设备

1. 设备
PC 机(安装 wave 编程软件、Keil C51 软件)、单片机实验板。
2. 工具及材料
工作对象：电工电子工具、电子元器件和辅助材料、仿真器、编程器。
工作工具：单片机控制电路原理图、实训指导书、项目任务单、工作记录单、项目检查单、各种电工仪表、常用电工工具和拆装工具、量具、相关电子手册。

硬件设计

主控模块采用 ATMEL 公司生产的 AT89S52 单片机。一键多功能按键模块与单片机的接口电路如图 13-1 所示。

软件设计

一、设计思想

在生活中，我们很容易通过姓名区别不同的人。同样，对于要通过一个按键来识别不同的功能，也可以给每个不同的功能模块用不同的 ID 号标识。这样，每按下一次按键，ID 的值是不相同的，所以单片机就很容易识别不同功能的身份了。

任务 13　一键多功能按键识别技术

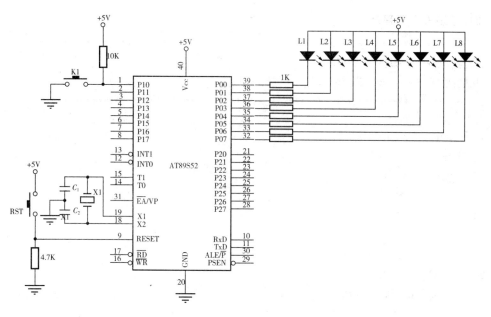

图 13-1　一键多功能按键模块与单片机的接口电路图

二、设计方法

从上面可以看出，L1 到 L4 发光二极管在每个时刻的闪烁的时间是受开关 SP1 来控制，我们给 L1 到 L4 闪烁的时段定义出不同的 ID 号。当 L1 在闪烁时，ID＝0；当 L2 在闪烁时，ID＝1；当 L3 在闪烁时，ID＝2；当 L4 在闪烁时，ID＝3；很显然，只要每次按下开关 K1 时，分别给出不同的 ID 号就能够完成上面的任务了。下面给出有关程序设计的框图，如图 13-2 所示。

三、程序流程图

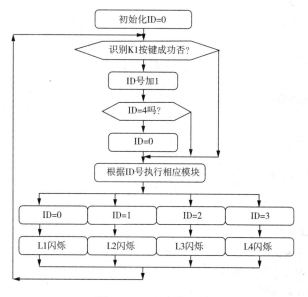

图 13-2　程序流程图

四、汇编语言源程序

```
;********************************************************************
;项目名称:一键多功能识别技术
;功能:根据开关 K1 按下的次数,L1—L4 轮流闪烁。
;********************************************************************
ID      EQU   30H
K1      BIT   P1.0
L1      BIT   P0.0
L2      BIT   P0.1
L3      BIT   P0.2
L4      BIT   P0.3
        ORG 0000H
        MOV  ID,#00H
START:JB  K1,REL
        LCALL DELAY10MS
        JB K1,REL
        INC ID
        MOV A,ID
        CJNE A,#04,REL
        MOV ID,#00H
REL:   JNB  K1,$
        MOV A,ID
        CJNE A,#00H,IS0
        CPL L1
        LCALL DELAY
        SJMP START
IS0:   CJNE  A,#01H,IS1
        CPL L2
        LCALL DELAY
        SJMP START
IS1:   CJNE  A,#02H,IS2
        CPL L3
        LCALL DELAY
        SJMP START
IS2:   CJNE  A,#03H,IS3
        CPL L4
        LCALL DELAY
        SJMP  START
IS3:   LJMP  START
        DELAY10MS:MOV  R6,#20
LOOP1:MOV  R7,#248
        DJNZ  R7,$
```

```
            DJNZ   R6,LOOP1
            RET
DELAY:MOV   R5,#20
LOOP2:LCALL  DELAY10MS
            DJNZ   R5,LOOP2
            RET
            END
```

五、C 语言源程序

```c
//*****************************************************************************
//项目名称:一键多功能识别技术
//功能:根据开关 K1 按下的次数,L1—L4 轮流闪烁。
//*****************************************************************************
#include<reg51.H>
unsigned char ID;
void delay10ms(void)
{
    unsigned char i,j;
    for(i=20;i>0;i--)
    for(j=248;j>0;j--);
}
void delay02s(void)
{
    unsigned char i;
    for(i=20;i>0;i--)
        {delay10ms();
        }
}
//主函数
void main(void)
{
        while(1)
        {  if(P1_0==0)
           {   delay10ms();
            if(P1_0==0)
            {
                ID++;
                if(ID==4)  {ID=0;}
                while(P1_0==0);
            }
         }
        switch(ID)
          { case 0:
```

```
            P0_0 = ~P0_0;
            delay02s();
            break;
        case 1:
            P0_1 = ~P0_1;
            delay02s();
            break;
        case 2:
            P0_2 = ~P0_2;
            delay02s();
            break;
        case 3:
            P0_3 = ~P0_3;
            delay02s();
            break;
        }
    }
}
```

实训考核

本课程改革传统的闭卷或开卷考核,而采用过程考核为主的多元化考核方式,考核分为理论考核、职业道德考核和技能考核三部分,各部分所占比例见表 13-1、表 13-2、表 13-3。

表 13-1 理论考核和职业素质考核形式及所占比例

序号	名称			比例	得分
一	理论考核	过程作业文件	个人自评	10%	
			组内互评	20%	
			小组互评	30%	
			老师评定	20%	
		课堂提问、解答		10%	
		项目汇报		10%	
		小计		100%	
二	职业素质	职业道德,工作作风		40%	
		小组沟通协作能力		40%	
		创新能力		20%	
		小计		100%	

表 13-2 技能考核内容及比例

姓名		班级		小组		总得分	
序号	考核项目	考核内容及要求		配分	评分标准	考核环节	得分
1	①安全文明生产 ②安全操作规范	着装规范		20%	现场考评	实施	
		安全用电					
		布线规范、合理					
		工具摆放整齐					
		工具及仪器仪表使用规范、摆放整齐					
		任务完成后,进行场地整理,保持场地清洁、有序					
2	实训态度	不迟到、早退、旷课		10%	现场考评	六步	
		实训过程认真负责					
		组内主动沟通、协作,小组间互助					
3	系统方案制定	工作流程正确合理		10%	现场考评	计划决策	
		方案合理					
		选用指令是否合理					
		电路图正确					
	编程能力	独立完成程序		10%	现场考评	决策	
		程序简单、可靠					
4	操作能力	正确输入程序并进行程序调试		20%	现场考评	实施	
		根据电路图正确接线					
		根据系统功能进行正确操作演示					
5	工艺	接线美观		10%	现场考评	实施	
		线路工作可靠					
6	实践效果	系统工作可靠		10%	现场考评	检查	
		满足工作要求					
		创新					
		按规定的时间完成项目					
7	汇报总结	工作总结,PPT 汇报		5%	现场考评	评估	
		填写自我检查表及反馈表					
8	技术文件制作整理	技术文件制作整理能力		5%	现场考评	评估	
		合计		100%			

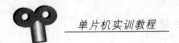

表 13-3　各部分考核占课程考核的比例

考核项目	理论考核	技能考核	职业素质考核	合　计
比　例	30%	50%	20%	100%
分　值	30	50	20	100
实际得分				

任务 14 4×4 矩阵键盘识别技术

实训任务

利用单片机的并行口 P1 接 4×4 矩阵键盘,以 P1.0～P1.3 作输入线,以 P1.4～P1.7 作输出线;在数码管上显示每个按键的"0～F"序号。对应的按键的序号排列如图 14-1 所示。

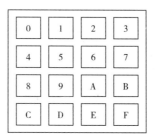

图 14-1 4×4 按键序号排列

实训设备

1. 设备

PC 机(安装 wave 编程软件、Keil C51 软件)、单片机实验板。

2. 工具及材料

工作对象:电工电子工具、电子元器件和辅助材料、仿真器、编程器。

工作工具:单片机控制电路原理图、实训指导书、项目任务单、工作记录单、项目检查单、各种电工仪表、常用电工工具和拆装工具、量具、相关电子手册。

硬件设计

主控模块采用 ATMEL 公司生产的 AT89S52 单片机。

本项目需要的按键数较多(为 16 个),通过前面的分析可知,需要采用矩阵式键盘。AT89S52 单片机的 P1 口用作键盘 I/O 口,键盘的列线接到 P1 口的低 4 位,键盘的行线接到 P1 口的高 4 位。列线 P1.0～P1.3 分别接有 4 个上拉电阻到正电源+5V,同时把列线 P1.0～P1.3 设置为输入线,行线 P1.4～P1.7 设置为输出线。4 根行线和 4 根列线形成 16 个相交点。矩阵键盘模块与单片机的接口电路如图 14-2 所示。

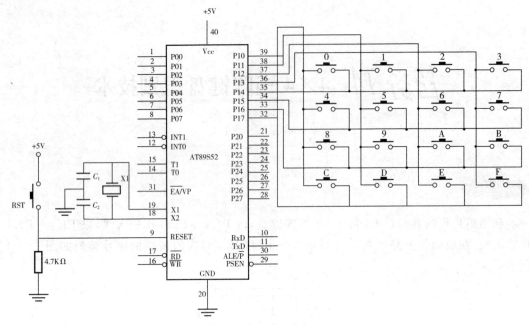

图 14-2 矩阵键盘模块与单片机的接口电路原理图

软件设计

一、4×4 矩阵键盘识别处理

每个按键有它的行值和列值,行值和列值的组合就是识别这个按键的编码。矩阵的行线和列线分别通过两并行接口和 CPU 通信。每个按键的状态同样需变成数字量"0"和"1",开关的一端(列线)通过电阻接 VCC,而接地是通过程序输出数字"0"实现的。键盘处理程序的任务是:确定有无键按下,判断哪一个键按下,键的功能是什么;还要消除按键在闭合或断开时的抖动。两个并行口中,一个输出扫描码,使按键逐行动态接地,另一个并行口输入按键状态,由行扫描值和回馈信号共同形成键编码而识别按键,通过软件查表,查出该键的功能。

二、程序设计

1. 主函数设计

主函数主要完成硬件初始化、程序调用等功能。

① 初始化

设置第一个数码管使能信号 P2.0 有效,P2.0=0,即 P2=0xfe。把控制数码管显示字形的 P0 口置为 0x07,系统上电显示"P"。

② 判断键盘中有无键按下

扫描第一行,将行线 P1.4 置低电平,即 P1=0xef,然后调用按键扫描函数检测列线的状态。

扫描第二行,将行线 P1.5 置低电平,即 P1=0xdf,然后调用按键扫描函数检测列线的状态。

扫描第三行,将行线 P1.6 置低电平,即 P1=0xbf,然后调用按键扫描函数检测列线的状态。

扫描第四行,将行线 P1.7 置低电平,即 P1=0x7f,然后调用按键扫描函数检测列线的状态。

在确定某根行线位置为低电平后,再逐行检测各列线的电平状态。若某列为低电平,则该列线与置为低电平的行线交叉处的按键就是闭合的按键。若所有列线均为高电平,则无按键按下。

行线与列线都扫描完成后,程序跳转到开始处进行下一轮行线与列线循环扫描。主函数设计流程图如图14-3所示。

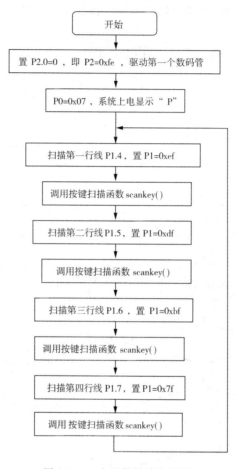

图 14-3 主函数设计流程图

2. 按键扫描函数设计

在置某根行线为低电平时,其他行线为高电平,然后确认有无键按下。判断方法是:只要有一列的电平为低,则表示键盘中有键被按下,而且闭合的键位于低电平线与4根行线相交叉的4个按键之中。

读P1口的值,并根据P1的值判断是否有键按下以及按下键的键号值。

若P1=0xee,则第一行线P1.4和第一列线P1.0为低电平,表明0号键被按下,并把键号值0作为存储数码管字形码的数组ledcode[]的下标,然后把ledcode[0]送给P0进行显示,最后程序退出。

若P1=0xed,则第一行线P1.4和第二列线P1.1为低电平,表明1号键被按下,并把键号值1作为存储数码管字形码的数组ledcode[]的下标,然后把ledcode[1]送给P0进行显示,最后程序退出。

若 P1=0xed,则第一行线 P1.4 和第二列线 P1.2 为低电平,表明 2 号键被按下,并把键号值 2 作为存储数码管字形码的数组 ledcode[]的下标,然后把 ledcode[2]送给 P0 进行显示,最后程序退出。

若 P1=0xed,则第一行线 P1.4 和第二列线 P1.3 为低电平,表明 3 号键被按下,并把键号值 3 作为存储数码管字形码的数组 ledcode[]的下标,然后把 ledcode[3]送给 P0 进行显示,最后程序退出。

其他按键的判断步骤与前面所述一致,不再赘述。按键扫描函数设计流程图如图 14-4 所示。

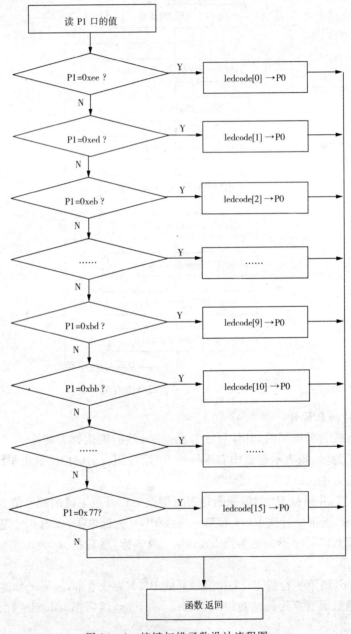

图 14-4　按键扫描函数设计流程图

三、C 语言源程序

```c
// ******************************************************************************
//项目名称:矩阵键盘识别技术
//功能:每按下一个按键(共 4x4 = 16 个),在数码管上显示相应的键号"0~F"
// ******************************************************************************
#include<reg51.h>           //包含"8051 寄存器定义"头文件
//变量定义:ledcode[ ]—存储数码管字形码
//程序中不变数据存放在片内的 CODE 存储区,以节省宝贵的 RAM
unsigned char code ledcode[ ] = {0xc0,0xf9,0xa4,0xb0,0x99,0x92,0x82,0xf8,0x80,0x90,0x88,
0x83,0xc6,0xa1,0x86,0x8e};
//函数声明:按键扫描函数 scankey()
void scankey(void);
//主函数
void main(void)
{
    E = 0;                  //取消 LCD 对 LED 的影响
    RW = 0;
    RS = 1;
    P2 = 0xfe;              //用第一个数码管显示,P2.0 = 0,即 P2 = 0xfe
    P0 = 0x07;              //系统上电,数码管显示"P"
    while(1)
    {
        P1 = 0xef;          //将第一行线 P1.4 置低电平,即 P1 = 0xef
        scankey();          //调用按键扫描函数 scankey()
        P1 = 0xdf;          //将第二行线 P1.5 置低电平,即 P1 = 0xdf
        scankey();          //调用按键扫描函数 scankey()
        P1 = 0Xbf;          //将第三行线 P1.6 置低电平,即 P1 = 0xbf
        scankey();          //调用按键扫描函数 scankey()
        P1 = 0x7f;          //将第四行线 P1.7 置低电平,即 P1 = 0x7f
        scankey();          //调用按键扫描函数 scankey()
    }
}
//按键扫描函数 scankey()
void scankey(void)
{
    switch(P1)
    {
        case 0xee:{ P0 = ledcode[0];}  break;       //按键"0"
        case 0xed:{ P0 = ledcode[1];}  break;       //按键"1"
        case 0xeb:{ P0 = ledcode[2];}  break;       //按键"2"
        case 0xe7:{ P0 = ledcode[3];}  break;       //按键"3"
        case 0xde:{ P0 = ledcode[4];}  break;       //按键"4"
```

```c
        case 0xdd:{ P0 = ledcode[5];}   break;          //按键"5"
        case 0xdb:{ P0 = ledcode[6];}   break;          //按键"6"
        case 0xd7:{ P0 = ledcode[7];}   break;          //按键"7"
        case 0xbe:{ P0 = ledcode[8];}   break;          //按键"8"
        case 0xbd:{ P0 = ledcode[9];}   break;          //按键"9"
        case 0xbb:{ P0 = ledcode[10];}  break;          //按键"a"
        case 0xb7:{ P0 = ledcode[11];}  break;          //按键"b"
        case 0x7e:{ P0 = ledcode[12];}  break;          //按键"c"
        case 0x7d:{ P0 = ledcode[13];}  break;          //按键"d"
        case 0x7b:{ P0 = ledcode[14];}  break;          //按键"e"
        case 0x77:{ P0 = ledcode[15];}  break;          //按键"f"
        default:                        break;          //函数退出返回
    }
}
```

四、汇编语言源程序

```asm
;********************************************************************
;项目名称:矩阵键盘识别技术
;功能:每按下一个按键(共 4X4 = 16 个按键),在数码管上显示相应的键号"0~F"
;********************************************************************
        ORG     0000H
        AJMP    MAIN            ;跳转到主程序
;功能:主程序
        ORG     0030H           ;主程序入口地址
MAIN:   CLR     P2.0            ;P2.0=0,用第一个 8 字数码管显示
        MOV     P0,#07H         ;系统上电显示"P"
LOOP:   MOV     P1,#0EFH        ;P1.4=0,扫描第一行
        ACALL   SCANKEY         ;调用按键扫描子程序
        MOV     P1,#0DFH        ;P1.5=0,扫描第二行
        ACALL   SCANKEY         ;调用按键扫描子程序
        MOV     P1,#0BFH        ;P1.6=0,扫描第三行
        ACALL   SCANKEY         ;调用按键扫描子程序
        MOV     P1,#07FH        ;P1.7=0,扫描第四行
        ACALL   SCANKEY         ;调用按键扫描子程序
        AJMP    LOOP            ;循环进行下一轮按键扫描
;功能:按键扫描子程序
SCANKEY: MOV    A,P1                    ;读 P1 口
        CJNE    A,#0EEH,NEXT1   ;判断列线 P1.0 是否为低电平
        MOV     A,#00H          ;0 号键被按下,把键号值 0 送给 A
        AJMP    SENDLED         ;跳转到数码管显示字形程序处
NEXT1:  CJNE    A,#0EDH,NEXT2   ;判断列线 P1.1 是否为低电平
        MOV     A,#01H          ;1 号键被按下,把键号值 0 送给 A
        AJMP    SENDLED         ;跳转到数码管显示字形程序处
```

NEXT2:	CJNE	A,#0EBH,NEXT3	;判断列线 P1.2 是否为低电平
	MOV	A,#02H	;2 号键被按下,把键号值 0 送给 A
	AJMP	SENDLED	;跳转到数码管显示字形程序处
NEXT3:	CJNE	A,#0E7H,NEXT4	;判断列线 P1.3 是否为低电平
	MOV	A,#03H	;3 号键被按下,把键号值 0 送给 A
	AJMP	SENDLED	;跳转到数码管显示字形程序处
NEXT4:	CJNE	A,#0DEH,NEXT5	;判断列线 P1.0 是否为低电平
	MOV	A,#04H	;4 号键被按下,把键号值 0 送给 A
	AJMP	SENDLED	;跳转到数码管显示字形程序处
NEXT5:	CJNE	A,#0DDH,NEXT6	;判断列线 P1.1 是否为低电平
	MOV	A,#05H	;5 号键被按下,把键号值 0 送给 A
	AJMP	SENDLED	;跳转到数码管显示字形程序处
NEXT6:	CJNE	A,#0DBH,NEXT7	;判断列线 P1.2 是否为低电平
	MOV	A,#06H	;6 号键被按下,把键号值 0 送给 A
	AJMP	SENDLED	;跳转到数码管显示字形程序处
NEXT7:	CJNE	A,#0D7H,NEXT8	;判断列线 P1.3 是否为低电平
	MOV	A,#07H	;7 号键被按下,把键号值 0 送给 A
	AJMP	SENDLED	;跳转到数码管显示字形程序处
NEXT8:	CJNE	A,#0BEH,NEXT9	;判断列线 P1.0 是否为低电平
	MOV	A,#08H	;8 号键被按下,把键号值 0 送给 A
	AJMP	SENDLED	;跳转到数码管显示字形程序处
NEXT9:	CJNE	A,#0BDH,NEXT_A	;判断列线 P1.1 是否为低电平
	MOV	A,#09H	;9 号键被按下,把键号值 0 送给 A
	AJMP	SENDLED	;跳转到数码管显示字形程序处
NEXT_A:	CJNE	A,#0BBH,NEXT_B	;判断列线 P1.2 是否为低电平
	MOV	A,#0AH	;A 号键被按下,把键号值 0 送给 A
	AJMP	SENDLED	;跳转到数码管显示字形程序处
NEXT_B:	CJNE	A,#0B7H,NEXT_C	;判断列线 P1.3 是否为低电平
	MOV	A,#0BH	;B 号键被按下,把键号值 0 送给 A
	AJMP	SENDLED	;跳转到数码管显示字形程序处
NEXT_C:	CJNE	A,#7EH,NEXT_D	;判断列线 P1.0 是否为低电平
	MOV	A,#0CH	;C 号键被按下,把键号值 0 送给 A
	AJMP	SENDLED	;跳转到数码管显示字形程序处
NEXT_D:	CJNE	A,#7DH,NEXT_E	;判断列线 P1.1 是否为低电平
	MOV	A,#0DH	;D 号键被按下,把键号值 0 送给 A
	AJMP	SENDLED	;跳转到数码管显示字形程序处
NEXT_E:	CJNE	A,#7BH,NEXT_F	;判断列线 P1.2 是否为低电平
	MOV	A,#0EH	;E 号键被按下,把键号值 0 送给 A
	AJMP	SENDLED	;跳转到数码管显示字形程序处
NEXT_F:	CJNE	A,#77H,SCAN_RE	;判断列线 P1.3 是否为低电平
	MOV	A,#0FH	;F 号键被按下,把键号值 0 送给 A
SENDLED:	MOV	DPTR,#LEDCODE	;把表头地址 LEDCODE 送给数据指针 DPTR
	MOVC	A,@A+DPTR	;查表

```
            MOV   P0,A              ;把查表所得的字形码送给P0,在数码管上显示
SCAN_RE:    RET                     ;子程序返回
;LED字段码表
LEDCODE:DB 0C0H, 0F9H, 0A4H, 0B0H, 99H, 92H, 82H, 0F8H, 80H, 90H, 88H, 83H, 0C6H, 0A1H, 86H,8EH
        END
```

实训考核

本课程改革传统的闭卷或开卷考核,而采用过程考核为主的多元化考核方式,考核分为理论考核、职业道德考核和技能考核三部分,各部分所占比例见表14-1、表14-2、表14-3。

表14-1 理论考核和职业素质考核形式及所占比例

序号	名称		比例	得分
一	理论考核	个人自评	10%	
		组内互评	20%	
		过程作业文件 小组互评	30%	
		老师评定	20%	
		课堂提问、解答	10%	
		项目汇报	10%	
		小计	100%	
二	职业素质	职业道德,工作作风	40%	
		小组沟通协作能力	40%	
		创新能力	20%	
		小计	100%	

表14-2 技能考核内容及比例

姓名		班级		小组		总得分	
序号	考核项目	考核内容及要求		配分	评分标准	考核环节	得分
1	①安全文明生产 ②安全操作规范	着装规范		20%	现场考评	实施	
		安全用电					
		布线规范、合理					
		工具摆放整齐					
		工具及仪器仪表使用规范、摆放整齐					
		任务完成后,进行场地整理,保持场地清洁、有序					

(续表)

2	实训态度	不迟到、早退、旷课	10%	现场考评	六步
		实训过程认真负责			
		组内主动沟通、协作,小组间互助			
3	系统方案制定	工作流程正确合理	10%	现场考评	计划决策
		方案合理			
		选用指令是否合理			
		电路图正确			
	编程能力	独立完成程序	10%	现场考评	决策
		程序简单、可靠			
4	操作能力	正确输入程序并进行程序调试	20%	现场考评	实施
		根据电路图正确接线			
		根据系统功能进行正确操作演示			
5	工艺	接线美观	10%	现场考评	实施
		线路工作可靠			
6	实践效果	系统工作可靠	10%	现场考评	检查
		满足工作要求			
		创新			
		按规定的时间完成项目			
7	汇报总结	工作总结,PPT汇报	5%	现场考评	评估
		填写自我检查表及反馈表			
8	技术文件制作整理	技术文件制作整理能力	5%	现场考评	评估
	合计		100%		

表 14-3 各部分考核占课程考核的比例

考核项目	理论考核	技能考核	职业素质考核	合 计
比 例	30%	50%	20%	100%
分 值	30	50	20	100
实际得分				

 任务 15 字符型 LCD 显示

实训任务

设计一个单片机控制的 LCD 显示字符系统。在 LCD 的第一行显示网站名：www.lzy.edu.cn，在第二行显示联系电话：0830—3150897。

实训设备

1. 设备

PC 机(安装 wave 编程软件、Keil C51 软件)、单片机实验板。

2. 工具及材料

工作对象：电工电子工具、电子元器件和辅助材料、仿真器、编程器。

工作工具：单片机控制电路原理图、实训指导书、项目任务单、工作记录单、项目检查单、各种电工仪表、常用电工工具和拆装工具、量具、相关电子手册。

硬件设计

主控模块采用 ATMEL 公司生产的 AT89S52 单片机，LCD 显示模块选用 1602 字符型 LCD 模块。LCD1602 显示模块与单片机 AT89S52 的接口电路如图 15-1 所示。

软件设计

一、LCD1602

1. 点阵型字符型 LCD 的接口特性

点阵字符型 LCD 是专门用于显示数字、字母、图形符号及少量自定义符号的液晶显示器。这类显示器把 LCD 控制器、点阵驱动器、字符存储器、显示体及少量的阻容元件等集成为一个液晶显示模块。鉴于字符型液晶显示模块目前在国际上已经规范化，其电特性及接口特性是统一的。因此，只要设计出一种型号的接口电路，在指令上稍加修改即可使用各种规格的字符型液晶显示块。我们这里主要介绍的是字符型液晶显示模块的控制器。

字符型液晶显示模块的控制器大多数为日立公司生产的 HD44780 及其兼容的控制电路，如 8ED1278（SEIKOEPSON）、KS0066（SAMSUNG）、NJU6408T 等。本节介绍 HD44780 的接口知识。

任务 15　字符型 LCD 显示

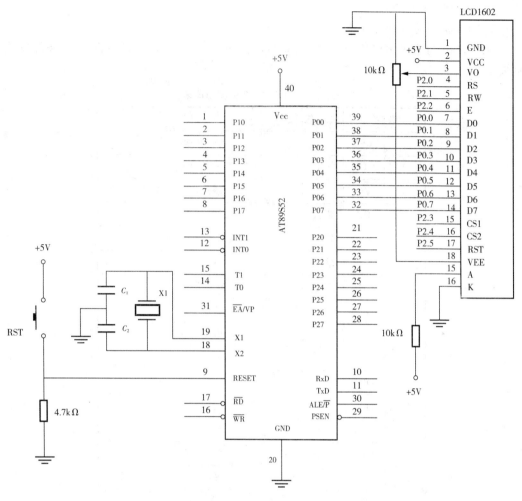

图 15-1　LCD 显示模块与单片机的接口电路原理图

2. 点阵字符型液晶显示模块的基本特点

字符型液晶显示模块的主要特点如下：

(1) 液晶显示屏是以若干 5×8 或 5×11 点阵块组成的显示字符群。每个点阵块为一个字符位，字符间距和行距都为一个点的宽度。

(2) 主控制电路为 HD44780 及其他公司的全兼容电路。因此从程序员角度来说，LCD 的显示接口与编程是面向 HD44780 的，只要了解 HD44780 的编程结构即可进行 LCD 的显示编程。

(3) 内部具有字符发生器 ROM，可显示 192 种字符(160 个 5×7 点阵符和 32 个 5×10 点阵字符)。

(4) 具有 64 字节的自定义字符 RAM，可以定义 8 个 5×8 点阵字符或 4 个 5×11 点阵字符。

(5) 具有 64 字节的数据显示 RAM，供进行显示编程时使用。

(6) 标准接口我，与 M68HC08 系列 MCU 容易接口。

(7) 模块结构紧凑、轻巧、装配容易。

(8) 单 +5V 电源供电(宽温型需要加 7V 驱动电源)。

(9)低功耗、高可靠性。

3. HD44780 的引脚与时序

HD44780 的外部接口信号一般有 16 条，与 MCU 的接口有 8 条数据线、3 条控制线。见表 15-1、15-2。

表 15-1　LCD1602 引脚功能表

引　脚	符　号	状　态	功　能
1	VSS		电源地
2	VDD		+5V 逻辑
3	VO		电源液晶驱动电源
4	RS	输入	寄存器选择;1:数据;0:指令
5	RW	输入	1:读操作;0:写操作
6	E	输入	使能信号
7	DB0	三态	数据总线(LSB)
8	DB1	三态	数据总线
9	DB2	三态	数据总线
10	DB3	三态	数据总线
11	DB4	三态	数据总线
12	DB5	三态	数据总线
13	DB6	三态	数据总线
14	DB7	三态	数据总线
15	A		背光 +
16	K		背光 -

表 15-2　HD44780 信号功能表

RS	R/W	E	功能
0	0	下降沿	写指令代码
0	1	高电平	读忙标志和 AC 值
1	0	下降沿	写数据
1	1	高电平	读数据

从 LCD1602 的时序图 15-2 可以清晰地看到数据写入的条件:寄存器 RS 为高电平，读/写标志 R/W 为低电平，建立地址，接下来使能信号 E 为高电平，数据被写入，当使能端 E 为下降沿的时候数据完全建立。从地址的建立、保持到数据的建立、保持的结束，整个过程需要的时间至少为 355ns。关于 MDLS 系列接口特性及电特性查点阵字符式液晶显示模块使用手册。

由于液晶的读数据用的地方不多，我们在此不讨论，有兴趣的可以查询这方面的资料。

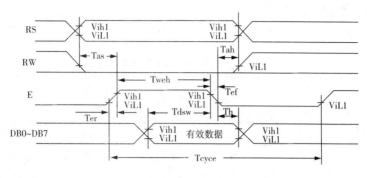

图 15-2 写操作时序图

4．HD44780 的编程结构

从编程角度看，HD44780 内部主要由指令寄存器(1R)、数据寄存器(DR)、忙标志(BF)、地址计数器(AC)、显示数据寄存器(DDRAM)、字符发生器 ROM(CGROM)、字符发生器 RAM(CGRAM)及时序发生电路构成。

(1) 指令寄存器(IR)

IR 用于 MCU 向 HD44780 写入指令码，IR 只能写入，不能读出。当 RS=0，R/W=0 时，数据线 DB7～DB0 上的数据写入指令寄存器 IR。

(2) 数据寄存器(DR)

DR 用于寄存数据。当 RS=1、R/W=0 时，数据线 DB7～DB0 上的数据写入数据寄存器 DR，同时 DR 的数据由内部操作自动写入 DDRAM 或 CGRAM。当 RS=1，R/W=1 时，内部操作将 DDRAM 或 CGRAM 送到 DR 中，通过 DR 送到数据总线 DB7～DB0 上。

(3) 忙标志(BF)

令 RS=0，R/W=1，在 E 信号高电平的作用下，BF 输出到总线的 DB7 上，MCU 可以读出判别。BF=1，表示组件正在进行内部操作，不能接受外部指令或数据。

(4) 地址计数器(AC)

AC 作为 DDRAM 或 CGRAM 的地址指针。如果地址码随指令写入 IR，则 IR 的地址码部分自动装入地址计数器 AC 之中，同时选择了相应的 DDRAM 或 CGRAM 单元。AC 具有自动加 1 和自动减 1 功能。当数据从 DR 送到 DD RAM(或 CG RAM)，AC 自动加 1。当数据 DDRAM(或 CGRAM)送到 DR，AC 自动减 1。当 RS=0、R/W =1 时，在 E 信号高电平的作用下，AC 的内容送到 DB7～DB0。

(5) 显示数据寄存器(DD RAM)

DD RAM 用于存储显示数据，共有 80 个字符码。对于不同的显示行数及每行字符个数，所使用的地址有所不同。

(6) 字符发生器 ROM(CG ROM)

CGROM 由 8 位字符码生成 5×7 点阵字符 160 种和 5×10 点阵字符 32 种。

(7) 字符发生器 RAM(CG RAM)

CGRAM 是为提供给用户自造特殊字符用的，它的容量仅为 64 字节，编址为 00～3FH。作为字符字模使用的仅是一个字节中的低 5 位，每个字节的高 3 位留给用户作为数据存储器使用。如果用户自定义字符由 5×7 点阵构成,可定义 8 个字符。

5. 字符型液晶指令集

字符型液晶指令集格式及功能如表 15-3 所示。

表 15-3 字符型液晶指令集格式及功能

指令名称	控制信号										运行时间 250KHZ	功 能
	RS	R/W	DB7	DB6	DB5	DB4	DB3	DB2	DB1	DB0		
清 屏	0	0	0	0	0	0	0	0	0	1	1.64ms	清 DDRAM 和 AC 的值
归 位	0	0	0	0	0	0	0	0	1	*	1.64ms	AC=0 光标、画面回 HOME 位
输入方式设置	0	0	0	0	0	0	0	1	I/D	S	40us	设置光标,画面移动方式
限制开关控制	0	0	0	0	0	0	1	D	C	B	40us	设置显示,光标及闪烁开/关
光标,画面位移	0	0	0	0	0	1	S/C	R/L	*	*	40us	光标,画面移动不影响 DDRAM
功能设置	0	0	0	0	1	DL	N	F	*	*	40us	工作方式设置(初始化指令)
CGRAM 地址设置	0	0	0	1	A5	A4	A3	A2	A1	A0	40us	设置 CGRAM 地址
DDRAM 地址设置	0	0	1	A6	A5	A4	A3	A2	A1	A0	40us	设置 DDRAM 地址
读 BF 及 AC 值	0	1	BF	AC6	AC5	AC4	AC3	AC2	AXC1	AC0	0us	读忙 BF 值地址计数器 AC 值
写数据	1	0	数 据								40us	数据写入 DDRAM/CGRAM
读数据	1	1	数 据								40us	DDRAM/CGRAM 数据读出

I/D=1:数据读/写操作后,AC 自动增 1	S/C=1:画面平移一个字符位	N=1:两行显示
I/D=0:数据读/写操作后,AC 自动减 1	S/C=0:光标平移一个字符位	N=0:一行显示
S=1:数据读/写操作,画面平移	R/L=1:右移	F=1:5*10 点阵字符
S=0:数据读/写操作,画面不动	R/L=0:左移	F=0:5*7 点阵字符
D:显示开关"1"-开;"0"-关	DL=1:8 位数据接口	BF=1:忙
C:光标开关"1"-开;"0"-关	DL=0:4 位数据接口	BF=0:准备好
B:闪烁开关"1"-开;"0"-关		

注:"*"表示任意值,在实际应用时一般认为是"0"。

(1)清屏指令使 DDRAM 的内容全部被清除,屏幕光标回原位,地址计数器 AC=0。运行时间(250KHz):1.64ms。

(2)归位指令使光标和光标所在位的字符回原点(屏幕的左上角)。但 DD RAM 单元内容不变。地址计数器 AC=0。运行时间(250KHz):1.64ms。

(3)输入方式设置。该指令设置光标、画面的移动方式。I/D=1:数据读写操作后,AC 自动增 1;I/D=0:数据读写操作后,AC 自动减 1;S=1:当数据写入 DDRAM,显示将全部左移(I/D=1)或全部右移(I/D=0),此时光标看上去未动,仅仅是显示内容移动,但从 DDRAM 中读取数据时,显示不移动;S=0:显示不移动,光标左移(I/D=1)或右移(I/D=0)。

(4)显示开关控制。该指令设置显示、光标及闪烁开、关。D:显示控制,D=1,开显示(Display ON),D=0,关显示(Display OFF);C:光标控制,C=1,开光标显示,C=0,关光标显示;B:闪烁控制,B=1,光标所指的字符同光标一起以 0.4s 交变闪烁,B=0,不闪烁。运行时间(250KHz):40μs。

(5)光标或画面移位。该指令使光标或画面在没有对 DD RAM 进行读写操作时被左移或右移,不影响 DD RAM。S/C=0、R/L=1,光标左移一个字符位,AC 自动减 1;S/C=0、R/L=1,光标右移一个字符位,AC 自动加 1;S/C=1,R/L=0,光标和画面一起左移一个字符位;S/C=I,R/L=1,光标和画面一起右移一个字符位。运行时间(250KHz):40μs。

(6)功能设置。该指令为工作方式设置命令(初始化命令)。对 HD44780 初始化时,需要设置数据接口位数(4 位或 8 位)、显示行数、点阵模式(5×7 或 5×10)。DL:设置数据接口位数,DL=I,8 位数据总线 DB7~DB0,DL=0,4 位数据总线 DB7~DB4,而 DB3~DB0 不用,在此方式下数据操作需两次完成;N:设置显示行数,N=1,2 行显示,N=0,1 行显示。F:设置点阵模式,F=0,5×7 点阵,F=1,5×10 点阵。运行时间(250KHz):40μs。

(7)CG RAM 地址设置。该指令设置 CGRAM 地址指针。A5~A0=00 0000~11 1111。地址码 A5~A0 被送入 AC 中,在此后,就可以将用户自定义的显示字符数据写入 CGRAM 或从 CGRAM 中读出。运行时间(250KHz):40μs。

(8)DD RAM 地址设置,该指令设置 DDRAM 地址指针。若是一行显示,地址码 A6~A0=00~4FH 有效,若是二行显示,首行址码 A6~A0=00~27H 有效,次行址码 A6~A0=40~67H 有效。在此后,就可以将显示字符码写入 DD RAM 或从 DD RAM 中读出。运行时间(250KHz):40μs。

(9)读忙标志 BF 和 AC 值,该指令读取 BF 及 AC。BF 为内部操作忙标志,BF=1,忙,BF=0,不忙。AC6~AC0 为地址计数器 AC 的值。当 BF=0 时,送到 DB6~DB0 的数据(AC6~AC0)有效。

(10)写数据到 DD RAM 或 CG RAM,该指令根据最近设置的地址性质,将数据写入 DD RAM 或 CGRAM 中。实际上,数据被直接写入 DR,再由内部操作写入地址指针所指的 DD RAM 或 CG RAM。运行时间(250KHz):4μs。

(11)读 DD RAM 或 CG RAM 数据,该指令根据最近设置的地址性质,从 DDRAM 或 CGRAM 读数

据到总线 DB7~DB0 上。运行时间(250KHz):40μs。

二、算法设计

把 LCD 需要显示的内容定义为字符数组,利用指向字符数组的指针对字符数组实现遍历访问。

三、程序设计

1. 主函数设计

主函数主要完成 LCD 显示字符数组变量定义、LCD 初始化、函数调用等功能。
(1)初始化。

定义 LCD 第一行显示字符数组变量 msg1[]和 LCD 第二行显示字符数组变量 msg2[]并赋初值,调用 LCD 初始化函数完成对 LCD 的初始化设置。

(2)字符显示。

完成对 LCD 初始化后,调用 LCD 字符显示函数显示第一行字符和第二行字符。
主函数设计流程图如图 15-3 所示。

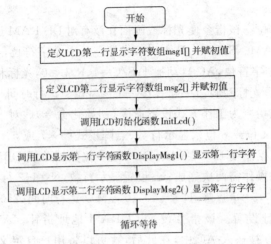

图 15-3 主函数设计流程图

2. 写入显示数据到 LCD 函数设计

当 LCD1602 的寄存器选择信号 RS 为 1 时,选择数据寄存器。当 LCD1602 的读写选择信号 R/W 为 0 时,进行写操作。把显示数据 ch 送至 P0 口(LCD 数据线 DB7~DB0)。当 LCD1602 的使能信号 E 至高电平后再过两个时钟周期至低电平,产生一个下降沿信号,往 LCD 写入显示数据。写入显示数据到 LCD 函数设计流程图如图 15-4 所示。

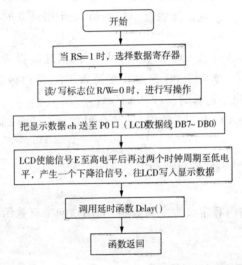

图 15-4 写入显示数据到 LCD 函数设计流程图

3. 写入指令数据到 LCD 函数设计

当 LCD1602 的寄存器选择信号 RS 为 0 时,选择指令寄存器。当 LCD1602 的读写选择线 R/W 为 0 时,进行写操作。把指令数据 ch 送至 P0 口(LCD 数据线 DB7~DB0)。当 LCD1602 的使能信号 E 置高电平后再过两个时钟周期置低电平,产生一个下降沿信号,往 LCD 写入指令代码。写入指令数据到 LCD 函数设计流程图如图 15-5 所示。

4. LCD 显示字符函数设计

首先定义变量 count,表示 LCD 显示字符数组下标索引值,并利用指向字符数组的指针 p 对字符数组进行访问。设置 LCD 的 DDRAM 地址,首行(0x80);次行(0xc0),并调用写入指令到 LCD 函数设置 DDRAM 地址指针。初始化 count 为 0，并判断 count 是否小于 16：

① 若 count 小于 16，调用写入显示数据到 LCD 函数，使数据显示在 LCD 上，再 count+1→count，并转向判断 count 是否小于 16。

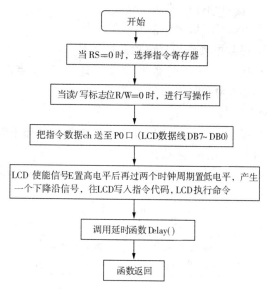

图 15-5　写入指令数据到 LCD 函数流程图

② 若 count 不小于 16，表明字符已经显示完成，函数返回。

LCD 显示字符函数设计流程图如图 15-6 所示。

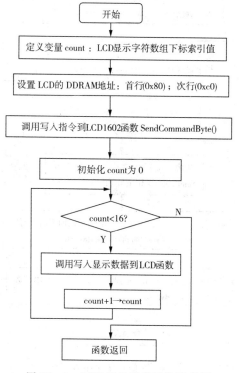

图 15-6　LCD 显示字符函数流程图

5. LCD 初始化函数设计

1602 字符型 LCD 的初始化过程为:

(1)延时 15ms,写指令 0x30(不检测忙信号)。

(2)延时 5ms,写指令 0x30(不检测忙信号)。

(3)延时 5ms,写指令 0x30(不检测忙信号)。

以后每次写指令、读/写数据操作均需要检测忙信号 BF。

(4)设置工作方式:写指令 0x38(8 位数据口,2 行显示,5*7 点阵)。

(5)设置显示状态:写指令 0x08(显示关)。

(6)清屏:写指令 0x01。

(7)设置输入方式:写指令 0x06(数据读写后 AC 自增 1,画面不动)。

(8)设置显示状态:写指令 0x0c(显示开)。

根据上述初始化过程,LCD 初始化函数设计流程图如图 15-7 所示。

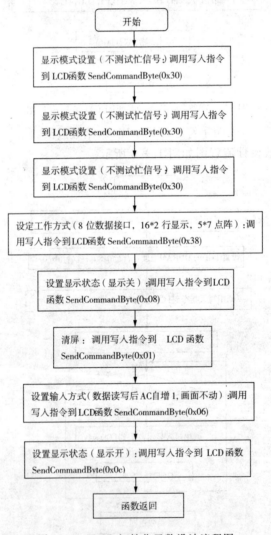

图 15-7 LCD 初始化函数设计流程图

四、C 语言源程序

```c
// ******************************************************************************
//项目名称:字符型 LCD 显示
//功能:在 1602 字符型液晶第一行显示网站名:www.lzy.edu.cn;
//在第二行显示联系电话:0830—3150897
// ******************************************************************************
#include<reg51.h>                      //包含"51 寄存器定义"头文件
//LCD1602 信号接口定义
sbit E = P2^2;                         //LCD 使能信号
sbit RW = P2^1;                        //读/写选择信号 R/W:0 为写入数据;1 为读出数据
sbit RS = P2^0;                        //数据/命令选择信号 R/S:0 为指令信号;1 为数据
                                       //  信号
//功能:变量定义
typedef unsigned char uchar;           //定义 unsigned char 为 uchar
//功能:函数声明
void Delay(unsigned int t);            //延时函数
void InitLcd(void);                    //LCD1602 初始化函数 InitLcd()
void SendCommandByte(unsigned char ch);//写入指令数据到 LCD1602 函数
void SendDataByte(unsigned char ch);   //写入显示数据到 LCD1602 函数
void DisplayMsg1(uchar *p);            //LCD 显示第一行字符函数 DisplayMsg1()
void DisplayMsg2(uchar *p);            //LCD 显示第二行字符函数 DisplayMsg2()
//主函数
main( )
{
    char msg1[16]= " AT89S51 DEMO";   //定义 LCD 第一行显示字符数组 msg1[]
    char msg2[16]= "www.mcuprog.com"; //定义 LCD 第二行显示字符数组 msg2[]
    InitLcd();                         //调用 LCD1602 初始化函数
    DisplayMsg1(msg1);                 //调用 LCD 显示第一行字符函数 DisplayMsg1()
    DisplayMsg2(msg2);                 //调用 LCD 显示第二行字符函数 DisplayMsg1()
    while(1);                          //循环等待
}
//功能:延时函数 Delay()
voidDelay(unsigned int t)              //delay
{
    for(;t! =0;t- -);
}
//功能:LCD1602 初始化函数 InitLcd( )
void InitLcd( )
{
    SendCommandByte(0x30);             //显示模式设置(不测试忙信号)共三次
    SendCommandByte(0x30);             //显示模式设置(不测试忙信号)共三次
    SendCommandByte(0x30);             //显示模式设置(不测试忙信号)共三次
```

```c
    SendCommandByte(0x38);        //设置工作方式:8位数据口,2行显示,5*7点阵
    SendCommandByte(0x08);        //设置显示状态:显示关
    SendCommandByte(0x01);        //清屏
    SendCommandByte(0x06);        //设置输入方式:数据读写后AC自增1,画面不动
    SendCommandByte(0x0c);        //设置显示状态:显示开
}
//******************************************************************************
//功能:写入指令数据到LCD1602函数 SendCommandByte()
//形参:ch
//******************************************************************************
void SendCommandByte(unsigned char ch)
{
    RS = 0;                       //数据/命令选择信号R/S:0为指令信号;1为数据
                                  //  信号
    RW = 0;                       //读/写选择信号R/W:0为写入数据;1为读出数据
    P0 = ch;                      //把指令数据ch送至P0口(LCD数据线DB7~DB0)
    E = 1;                        //LCD使能信号E置高电平
    Delay(1);                     //延时0.40us
    E = 0;                        //LCD使能信号E置低电平
    Delay(100);//delay40us
}

//******************************************************************************
//功能:写入显示数据到LCD1602函数 SendDataByte()
//形参:ch
//******************************************************************************
void SendDataByte(unsigned char ch)
{
    RS = 1;                       //数据/命令选择信号R/S:0为指令信号;1为数据
                                  //  信号
    RW = 0;                       //读/写选择信号R/W:0为写入数据;1为读出数据
    P0 = ch;                      //把显示数据ch送至P0口(LCD数据线DB7~DB0)
    E = 1;                        //LCD使能信号E置高电平
    Delay(1);                     //延时0.40us
    E = 0;                        //LCD使能信号E置低电平
    Delay(100);//delay 40us
}
//******************************************************************************
//功能:LCD显示第一行字符函数 DisplayMsg1()
//形参:字符指针p
//******************************************************************************
void DisplayMsg1(uchar *p)
{
    unsigned char count;          //定义LCD显示字符数组下标索引值
```

```c
    SendCommandByte(0x80);              //设置DDRAM地址:首行地址码00H—27H
    for(count=0;count<16;count++)       //利用for循环依次显示第一行字符
    { SendDataByte(*p++);}              //调用写入显示数据到LCD1602函数
}
//**************************************************************
//功能:LCD显示第二行字符函数DisplayMsg2()
//形参:字符指针p
//**************************************************************
void DisplayMsg2(uchar *p)
{
    unsigned char count;                //定义LCD显示字符数组下标索引值
    SendCommandByte(0xc0);              //设置DDRAM地址:次行地址码40H—67H
    for(count=0;count<16;count++)       //利用for循环依次显示第二行字符
    { SendDataByte(*p++);  }            //调用写入显示数据到LCD1602函数
}
```

五、汇编源程序

```
;**************************************************************
;项目名称:字符型LCD显示
;功能:在1602字符型液晶第一行显示网站名:www.lzy.edu.cn
;在第二行显示联系电话:0830—3150897
;**************************************************************
;LCD1602信号接口定义
E           BIT     P2.2        ;LCD使能信号
RW          BIT     P2.1        ;LCD读/写选择信号R/W:0为写入数据,1为读出数据
RS          BIT     P2.0        ;LCD数据/命令选择信号R/S:0为指令,1为数据
LCDPORT     EQU     P0          ;LCD1602数据线DB7~DB0
CMD_BYTE    EQU     2EH         ;写入指令数据到LCD1602入口参数
DAT_BYTE    EQU     2FH         ;写入显示数据到LCD1602入口参数
;功能:主程序
        ORG     0000H
        AJMP    MAIN
        ORG     0050H
MAIN:   MOV     SP,#60H         ;设置堆栈指针
        LCALL   INITLCD         ;调用LCD初始化子程序INITLCD
        LCALL   DISPMSG1        ;调用液晶字符显示程序DISPMSG1
        LCALL   DISPMSG2        ;调用液晶字符显示程序DISPMSG2
        SJMP    $
;**************************************************************
;LCD1602要用到的一些子程序
;**************************************************************
;功能:写入指令数据到LCD1602子程序
WRITE_CMD:  CLR     RS          ;当RS=0,RW=0时,写指令
```

```
                CLR   RW
                MOV   A,CMD_BYTE          ;把写命令入口参数 CMD_BYTE 传给 A
                MOV   LCDPORT, A          ;LCDPORT 为液晶数据线 DB7～DB0
                SETB  E                   ;打开液晶使能信号 E
                NOP
                NOP
                CLR   E                   ;关闭液晶使能信号 E
                LCALL DELAY               ;调用延时子程序
                RET
;功能:写入显示数据到 LCD1602 子程序
WRITE_DAT:      SETB  RS                  ;当 RS＝1,RW＝0 时,写入数据到 LCD
                CLR   RW
                MOV   A,DAT_BYTE          ;把写数据入口参数 DAT_BYTE 传给 A
                MOV   LCDPORT,A           ;LCDPORT 为液晶数据线 DB7～DB0
                SETB  E                   ;打开 LCD 使能信号 E
                NOP
                NOP
                CLR   E                   ;关闭 LCD 使能信号 E
                LCALL DELAY               ;调用延时子程序
                RET                       ;写入显示数据到 LCD1602 子程序返回
;函数功能:LCD 显示初始化子程序
INITLCD:   MOV   CMD_BYTE, ＃30H
           LCALL WRITE_CMD                ;调用写入指令数据到 LCD1602 子程序
           MOV   CMD_BYTE, ＃30H
           LCALL WRITE_CMD                ;调用写入指令数据到 LCD1602 子程序
           MOV   CMD_BYTE, ＃30H
           LCALL WRITE_CMD                ;调用写入指令数据到 LCD1602 子程序
           MOVCMD_BYTE, ＃38H              ;设定工作方式
           LCALL WRITE_CMD                ;调用写入指令数据到 LCD1602 子程序
           MOVCMD_BYTE, ＃0CH              ;显示状态设置
           LCALL WRITE_CMD                ;调用写入指令数据到 LCD1602 子程序
           MOVCMD_BYTE, ＃01H              ;DB7～DB0＝01H 为清屏
           LCALL WRITE_CMD                ;调用写入指令数据到 LCD1602 子程序
           MOVCMD_BYTE, ＃06H              ;输入方式设置
           LCALL WRITE_CMD                ;调用写入指令数据到 LCD1602 子程序
           RET
;功能:LCD 显示第一行字符子程序(在第一行显示表格 TB1 的内容)
DISPMSG1:  MOV   CMD_BYTE, ＃80H           ;设置 DDRAM 的地址
           LCALL WRITE_CMD                ;调用写命令子程序
           MOV   R7,＃10H                  ;R7 为显示数据个数
           MOV   R6,＃00H                  ;R6 表头地址
```

```
                MOV   DPTR,#TAB1
DISPMSG1_1:  MOV   A,R6
             MOVC A,@A+DPTR          ;查表
             MOV   DAT_BYTE,A         ;从A把数据传给写入数据入口参数 DAT_BYTE
             LCALL  WRITE_DAT          ;调用写数据子程序
             INC  R6                  ;下一个数据的地址
             DJNZ  R7,DISPMSG1_1       ;判断是否显示完毕
             RET                      ;子程序返回
;功能:LCD 显示第二行字符子程序(在第二行显示表格 TB2 的内容)
DISPMSG2:   MOV   CMD_BYTE,#0C0H      ;设置 DDRAM 的地址
            LCALL  WRITE_CMD           ;调用写入指令数据到 LCD1602 子程序
            MOV   R7,#10H
            MOV   R6,#00H
            MOV   DPTR,#TAB2
DISPMSG2_1:MOV   A,R6
            MOVC  A,@A+DPTR
            MOV   DAT_BYTE,A
            LCALL  WRITE_DAT           ;调用写入显示数据到 LCD1602 子程序
            INC   R6
            DJNZ   R7,DISPMSG2_1
            RET                      ;子程序返回
;功能:延时子程序
DELAY:   MOV  R5,#0A0H
DELAY1:  NOP
         DJNZ  R5,DELAY1
         RET                        ;延时子程序返回
;LCD1602 要显示的内容
         ORG   0200H
TAB1:   DB   " www.lzy.edu.cn"
TAB2:   DB   "0830—3150897"
END
```

实训考核

本课程改革传统的闭卷或开卷考核,而采用过程考核为主的多元化考核方式,考核分为理论考核、职业道德考核和技能考核三部分,各部分所占比例见表15-4、表15-5、表15-6。

表15-4　理论考核和职业素质考核形式及所占比例

序号	名称		比例	得分
一	理论考核	过程作业文件 个人自评	10%	
		过程作业文件 组内互评	20%	
		过程作业文件 小组互评	30%	
		过程作业文件 老师评定	20%	
		课堂提问、解答	10%	
		项目汇报	10%	
		小计	100%	
二	职业素质	职业道德,工作作风	40%	
		小组沟通协作能力	40%	
		创新能力	20%	
		小计	100%	

表15-5　技能考核内容及比例

姓名		班级		小组		总得分	
序号	考核项目	考核内容及要求		配分	评分标准	考核环节	得分
1	①安全文明生产 ②安全操作规范	着装规范		20%	现场考评	实施	
		安全用电					
		布线规范、合理					
		工具摆放整齐					
		工具及仪器仪表使用规范、摆放整齐					
		任务完成后,进行场地整理,保持场地清洁、有序					
2	实训态度	不迟到、早退、旷课		10%	现场考评	六步	
		实训过程认真负责					
		组内主动沟通、协作,小组间互助					
3	系统方案制定	工作流程正确合理		10%	现场考评	计划决策	
		方案合理					
		选用指令是否合理					
		电路图正确					
	编程能力	独立完成程序		10%	现场考评	决策	
		程序简单、可靠					

(续表)

4	操作能力	正确输入程序并进行程序调试	20%	现场考评	实施	
		根据电路图正确接线				
		根据系统功能进行正确操作演示				
5	工艺	接线美观	10%	现场考评	实施	
		线路工作可靠				
6	实践效果	系统工作可靠	10%	现场考评	检查	
		满足工作要求				
		创新				
		按规定的时间完成项目				
7	汇报总结	工作总结,PPT汇报	5%	现场考评	评估	
		填写自我检查表及反馈表				
8	技术文件制作整理	技术文件制作整理能力	5%	现场考评	评估	
	合计		100%			

表15-6 各部分考核占课程考核的比例

考核项目	理论考核	技能考核	职业素质考核	合 计
比 例	30%	50%	20%	100%
分 值	30	50	20	100
实际得分				

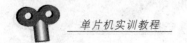

任务 16 按键变量加减 LCD 显示

实训任务

利用按键 K1（P1.0）、按键 K2（P1.1）来对变量进行加减，并且将结果显示在 LCD1602 上。

实训设备

1. 设备

PC 机（安装 wave 编程软件、Keil C51 软件）、单片机实验板。

2. 工具及材料

工作对象：电工电子工具、电子元器件和辅助材料、仿真器、编程器。

工作工具：单片机控制电路原理图、实训指导书、项目任务单、工作记录单、项目检查单、各种电工仪表、常用电工工具和拆装工具、量具、相关电子手册。

硬件设计

主控模块采用 ATMEL 公司生产的 AT89S52 单片机。按键 K1（P1.0）按下时，变量加 1；按键 K2（P1.1）按下时，变量减 1。按键模块与单片机的接口电路如图 16-1 所示。

软件设计

一、汇编语言源程序

```
;********************************************************************
;项目名称:按键变量加减 LCD 显示
;功能:利用按键 K1(P1.0)、按键 K2(P1.1)来对变量进行加减,并且将结果显示在 LCD1602 上。
;********************************************************************
;LCD 口线的定义
E         BIT   P2.2
RW        BIT   P2.1
RS        BIT   P2.0
LCDPORT   EQU   P0
CMD_BYTE  EQU   2EH
DAT_BYTE  EQU   2FH
```

· 118 ·

任务 16 按键变量加减 LCD 显示

```
KEY       BIT BIT     00H        ;有按键按下标志
DPBL      EQU         30H        ;当前显示的变量(20 到 200 间)
LEDBAI    EQU         31H        ;显示的百位
LEDSHI    EQU         32H        ;显示的十位
LEDGE     EQU         33H        ;显示的个位
LEDSM     EQU         34H        ;现在扫到第几个 LED
KEYTIME   EQU         35H        ;20MS 扫一次按键
          ORG 0000H
          AJMP MAIN
          ORG 000BH
          AJMP TIME0_1
;主程序
          ORG 0030H
MAIN:     MOV 20H,#00H
          LCALL INITLCD
          MOV DPBL,#20
          MOV KEYTIME,#04H
          MOV CMD_BYTE,#80H      ;设置 DDRAM 的地址
          LCALL WRITE_CMD
          LCALL DELAY0
          MOV DAT_BYTE,#"S"
          LCALL WRITE_DAT
          MOV DAT_BYTE,#"U"
          LCALL WRITE_DAT
          MOV DAT_BYTE,#"M"
          LCALL WRITE_DAT
          MOV DAT_BYTE,#":"
          LCALL WRITE_DAT
          MOV TH0,#0EEH          ;16 位的定时器,定时 5MS
          MOV TL0,#00H
          MOV TMOD,#01H
          SETB TR0
          MOV IE,#82H            ;定时器 0 中断
          SJMP $
          ORG 0100H
TIME0_1:  MOV TH0,#0EEH
          MOV TL0,#00H
          DJNZ KEYTIME,TIME0_RE
          MOV KEYTIME,#04H       ;每 20MS 进行一次按键扫描
          MOV P1,#0FFH
          LCALL KEYSCAN
          LCALL JSDPBL
          LCALL DISPDPBL
```

```
TIME0_RE:      RETI
;按键扫描
KEYSCAN:       NOP
NEXT_UP:       JB   P1.0,NEXT_DN        ;+键
               JB KEYBIT,SCAN_RE
               SETB KEYBIT
               INC DPBL
               MOV A,DPBL
               CJNE  A,#201,SCAN_RE     ;大于200反回20
               MOV DPBL,#20
               AJMP   SCAN_RE
NEXT_DN:       JB P1.1,NEXT_NC          ;-键
               JB KEYBIT,SCAN_RE
               SETB  KEYBIT
               DEC DPBL
               MOV A,DPBL
               CJNE  A,#19,SCAN_RE      ;小于20反回200
               MOV DPBL,#200
               AJMP SCAN_RE
NEXT_NC:       CLR KEYBIT               ;清标志用于等待下一次按键
               SCAN_RE:MOV P1,#0FFH
               RET
;LCD1602要用到的一些子程序
;写命令(入口参数 CMD_BYTE)
WRITE_CMD:     CLR RS
               CLR RW
               MOV A,CMD_BYTE
               MOV LCDPORT,A
               SETB E
               NOP
               NOP
               CLR E
               LCALL DELAY0
               RET
;写显示数据(入口参数 DAT_BYTE)
WRITE_DAT:     SETB RS
               CLR RW
               MOV A,DAT_BYTE
               MOV LCDPORT,A
               SETB  E
               NOP
               NOP
               CLR  E
```

```
                LCALL DELAY0
                RET
;LCD 显示初始化
INITLCD:        MOV CMD_BYTE,#30H
                LCALL WRITE_CMD
                MOV CMD_BYTE,#30H
                LCALL WRITE_CMD
                MOV CMD_BYTE,#30H
                LCALL WRITE_CMD
                MOV CMD_BYTE,#38H        ;设定工作方式
                LCALL WRITE_CMD
                MOV CMD_BYTE,#0CH        ;显示状态设置
                LCALL WRITE_CMD
                MOV CMD_BYTE,#01H        ;清屏
                LCALL WRITE_CMD
                MOV CMD_BYTE,#06H        ;输入方式设置
                LCALL WRITE_CMD
                RET
;延时子程序
DELAY0:         MOV  R5,#0A0H
DELAY1:         NOP
                DJNZ R5,DELAY1
                RET
;显示变量
DISPDPBL:       MOV CMD_BYTE,#85H        ;设置 DDRAM 的地址
                LCALL WRITE_CMD
                MOV DAT_BYTE,LEDBAI
                LCALL WRITE_DAT
                MOV DAT_BYTE,#"."
                LCALL WRITE_DAT
                MOV DAT_BYTE,LEDSHI
                LCALL WRITE_DAT
                MOV DAT_BYTE,LEDGE
                LCALL WRITE_DAT
                RET
;计算显示的值
JSDPBL:         MOV A,DPBL
                MOV B,#64H
                DIV AB
                OR L A,#30H
                MOV LEDBAI,A
                MOV A,B
                MOV B,#0AH
```

```
DIV   AB
ORLA,#30H
MOVLEDSHI,A
MOVA,B
ORLA,#30H
MOVLEDGE,A
RET
END
```

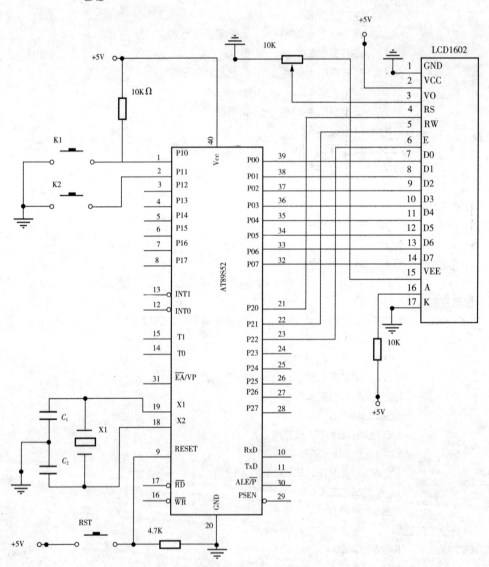

图 16-1 按键模块与单片机的接口电路图

任务16 按键变量加减 LCD 显示

二、C 语言源程序

```c
// ******************************************************************************
//项目名称:按键变量加减 LCD 显示
//功能:利用按键 K1(P1.0)、按键 K2(P1.1)来对变量进行加减,并且将结果显示在 LCD1602 上。
// ******************************************************************************
#include<reg51.h>
//LCD 的口线定义
sbit E = P2^2;
sbit RW = P2^1;
sbit RS = P2^0;
unsigned char dpbl,ledbai,ledshi,ledge,keytime;
unsigned char bdata myflag;
sbit keybit = myflag^0;
void time0(void);
void keyscan(void);
void Delay(unsigned int t);
void SendCommandByte(unsigned char ch);
void SendDataByte(unsigned char ch);
void InitLcd(void);
void dispdpbl(void);
//主函数
void main(void)
{
        InitLcd();
        myflag = 0x00;
        dpbl = 20;
        keytime = 4;             //20MS 扫一次按键
        SendCommandByte(0x80);
        Delay(2);
        SendDataByte('S');
        SendDataByte('U');
        SendDataByte('M');
        SendDataByte(':');
        TH0 = 0xee;              //定时 5MS
        TL0 = 0x00;
        TMOD = 0x01;
        TR0 = 1;
        IE = 0x82;
        while(1)
        { }
}
//定时器 T0 中断函数
```

```c
void time0(void) interrupt 1
{
    TH0 = 0xee;
    TL0 = 0x00;
    if(keytime - - = = 0)
    {keytime = 0x04;P1 = 0xff;keyscan();}
    ledbai = (dpbl/100)|0x30;
    ledshi = ((dpbl%100)/10)|0x30;
    ledge = (dpbl%10)|0x30;
    dispdpbl();
    }
}
//按键扫描函数
void keyscan(void)
{
    if(P1 = = 0xfe)
    {
        if(keybit = = 0)
        {
            keybit = 1;
            dpbl + + ;
            if(dpbl>200)  {dpbl = 20;}
        }
    }
    else
    {
        if(P1 = = 0xfd)
        {
            if(keybit = = 0)
            {
                keybit = 1;
                dpbl - - ;
                if(dpbl<20)  {dpbl = 200;}
            }
        }
        else
        {
            if(P1 = = 0xff)  {keybit = 0;}
        }
    }
    P1 = 0xff;
}
//延时函数
```

```c
void Delay(unsigned int t)        // delay 40us
{
    for(;t! =0;t--);
}
//LCD1602 写命令函数
void SendCommandByte(unsigned char ch)
{
    RS=0;
    RW=0;
    P0=ch;
    E=1;
    Delay(1);
    E=0;
    Delay(100);              //delay 40us
}
//LCD1602 写数据函数
void SendDataByte(unsigned char ch)
{   RS=1;
    RW=0;
    P0=ch;
    E=1;
        Delay(1);
    E=0;
    Delay(100);//delay 40us
}
//LCD1602 初始化
void InitLcd(void)
{
    SendCommandByte(0x30);
    SendCommandByte(0x30);
    SendCommandByte(0x30);
    SendCommandByte(0x38); //设置工作方式
    SendCommandByte(0x0c); //显示状态设置
    SendCommandByte(0x01); //清屏
    SendCommandByte(0x06); //输入方式设置
}
//变量值显示函数
void dispdpbl(void)
{
    SendCommandByte(0x85);
    SendDataByte(ledbai);
    SendDataByte('.');
    SendDataByte(ledshi);
```

```
    SendDataByte(ledge);
}
```

实训考核

本课程改革传统的闭卷或开卷考核,而采用过程考核为主的多元化考核方式,考核分为理论考核、职业道德考核和技能考核三部分,各部分所占比例见表 16-1、表 16-2、表 16-3。

表 16-1 理论考核和职业素质考核形式及所占比例

序号	名称		比例	得分
一	理论考核	过程作业文件 个人自评	10%	
		过程作业文件 组内互评	20%	
		过程作业文件 小组互评	30%	
		过程作业文件 老师评定	20%	
		课堂提问、解答	10%	
		项目汇报	10%	
		小计	100%	
二	职业素质	职业道德,工作作风	40%	
		小组沟通协作能力	40%	
		创新能力	20%	
		小计	100%	

表 16-2 技能考核内容及比例

姓名		班级		小组		总得分	
序号	考核项目	考核内容及要求		配分	评分标准	考核环节	得分
1	①安全文明生产 ②安全操作规范	着装规范		20%	现场考评	实施	
		安全用电					
		布线规范、合理					
		工具摆放整齐					
		工具及仪器仪表使用规范、摆放整齐					
		任务完成后,进行场地整理,保持场地清洁、有序					
2	实训态度	不迟到、早退、旷课		10%	现场考评	六步	
		实训过程认真负责					
		组内主动沟通、协作,小组间互助					

(续表)

3	系统方案制定	工作流程正确合理	10%	现场考评	计划决策
		方案合理			
		选用指令是否合理			
		电路图正确			
	编程能力	独立完成程序	10%	现场考评	决策
		程序简单、可靠			
4	操作能力	正确输入程序并进行程序调试	20%	现场考评	实施
		根据电路图正确接线			
		根据系统功能进行正确操作演示			
5	工艺	接线美观	10%	现场考评	实施
		线路工作可靠			
6	实践效果	系统工作可靠	10%	现场考评	检查
		满足工作要求			
		创新			
		按规定的时间完成项目			
7	汇报总结	工作总结，PPT汇报	5%	现场考评	评估
		填写自我检查表及反馈表			
8	技术文件制作整理	技术文件制作整理能力	5%	现场考评	评估
	合计		100%		

表16-3 各部分考核占课程考核的比例

考核项目	理论考核	技能考核	职业素质考核	合计
比例	30%	50%	20%	100%
分值	30	50	20	100
实际得分				

任务17 报警产生器

实训任务

设计一个单片机控制的报警器系统。当按钮开关 SB(SB 接 P3.3,外部中断 1 请求输入端)进行控制按下时,单片机输出 1kHz 和 500Hz 的音频信号驱动蜂鸣器发出报警信号,要求 1kHz 信号响 100ms,500Hz 信号响 200ms,交替进行。

实训设备

1. 设备
PC 机(安装 wave 编程软件、Keil C51 软件)、单片机实验板。
2. 工具及材料
工作对象:电工电子工具、电子元器件和辅助材料、仿真器、编程器。
工作工具:单片机控制电路原理图、实训指导书、项目任务单、工作记录单、项目检查单、各种电工仪表、常用电工工具和拆装工具、量具、相关电子手册。

硬件设计

主控模块采用 ATMEL 公司生产的 AT89S52 单片机。
按键 SB 的一端接地 VSS,另一端接 P3 口的 P3.3(INT1,外部中断 1 请求输入端),如图 4-4 所示。P3 口内部接有上拉电阻,所以在按键 SB 按下之前,端口 P3.3 保持在高电平;当按键 SB 按下时,端口 P3.3 通过按键 SB 接到 VSS,这个时候就是低电平。所以,通过检测端口的状态变化,可以判断按键 SB 是否按下。

在本项目中,选用无源电磁式蜂鸣器来实现报警发声。蜂鸣器的正极接到 V_{cc}(+5V)电源上面,蜂鸣器的负极接到三极管的发射极 E,三极管的基级 B 经过限流电阻 R_1 后由单片机的 P3.5 引脚控制。当 P3.5 输出高电平时,三极管 C8550 截止,没有电流流过线圈,蜂鸣器不发声;当 P3.5 输出低电平时,三极管 C8550 导通,蜂鸣器的电流形成回路,发出声音。因此,可以通过程序控制 P3.5 脚的电平来控制蜂鸣器是否发出声音。报警器与单片机的接口电路如图 17-1 所示。

任务 17 按键变量加减 LCD 显示

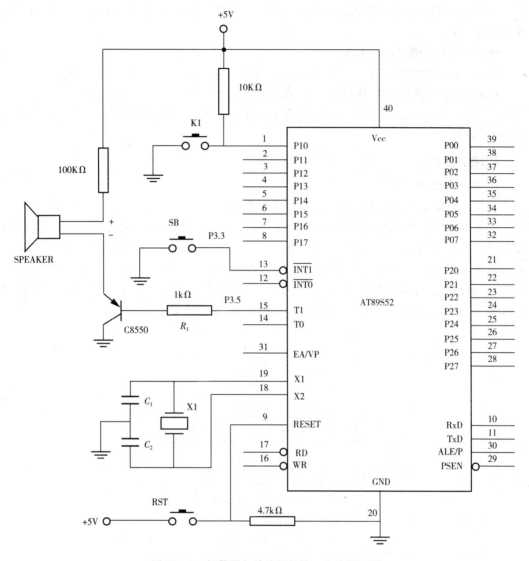

图 17-1 报警器与单片机的接口电路原理图

软件设计

一、算法设计

当按键 SB 未按下时,P3.3 口线(INT1,外部中断 1 请求输入端)为高电平;当按键 SB 按下时,P3.3 口线为低电平。单片机在相继的两个周期采样过程中,一个机器周期采样到该引脚为高电平,接着的下一个机器周期采样到该引脚为低电平时,则使外部中断 1 的中断请求标志 IE1 置 1,产生中断。

改变单片机 P3.5 引脚输出波形的频率,就可以调整控制蜂鸣器的音调,产生各种不同音色、音调的声音。另外,改变 P3.5 输出电平的高低电平占空比,则可以控制蜂鸣器的声音大小。

在中断服务函数中,调用延时函数并对P3.5引脚电平取反,来实现特定频率的报警音频信号的产生。

报警音频信号产生的方法如下:

500Hz信号的周期为1/500Hz=2ms,信号电平为每2ms/2=1ms取反1次。

1kHz的信号周期为1/1kHz=1ms,信号电平每1ms/2=500us取反1次。

1ms正好为500us的2倍,可以利用延时500us的延时函数来实现延时,1ms正好调用2次延时函数。

二、程序设计

1. 主函数设计

主程序主要完成对相关特殊功能寄存器初始化设置。

堆栈指针初始化,栈底设置为0x60。开放总中断,设置EA位;允许/INT1外部中断,设置EX1位。设置IT1位,使外部中断源/INT1工作于边沿触发方式下。外部中断/INT1对应的外部中断源是/INT1引脚,即P3.3,当外部中断1允许且为边沿触发方式时,只要在P3.3引脚上出现负的下降沿,外部中断1的标志位IE1就被置位,CPU将在下一个机器周期的S1状态时响应该中断。主程序设计流程图如图17-2所示。

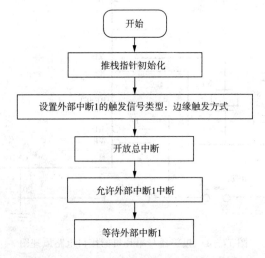

图17-2 主函数设计流程图

2. 外部中断1服务函数设计

CPU响应了外部中断1的中断请求后转至中断服务函数执行。其主要功能就是将P3.5的值取反、延时,再取反、再延时,从而实现P3.5口线交替输出1kHz和500Hz的音频信号来驱动蜂鸣器报警。

外部中断1服务函数设计流程如图17-3所示。

任务 17　按键变量加减 LCD 显示

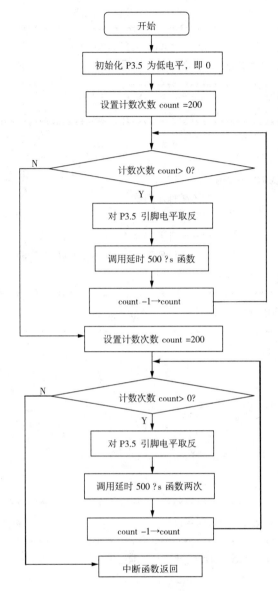

图 17-3　外部中断 1 服务函数设计流程图

三、C 语言源程序

```
//************************************************************************
//项目名称:报警器设计
//功能:按钮开关 SB 按下时,单片机输出 1kHz 和 500Hz 报警信号
//************************************************************************
#include<AT89X51.H>        //包含"51 寄存器定义"头文件
#include<INTRINS.H>        //包含指示编译器产生嵌入式固有代码的程序原型
//功能:变量定义
unsigned char count;       //设置计数次数
//************************************************************************
```

//功能:函数声明
// ***
```c
void delay500us(void);            //延时 500us 函数 delay500us()
void ExternalInterrupt1(void);    //外部中断 1 服务函数
```
// ***
//功能:主程序
// ***
```c
void main(void)
{
    SP = 0x60;     //堆栈指针初始化
    IT1 = 1;                     //边缘触发方式
    EA = 1;                      //开放总中断
    EX1 = 1;                     //开放外部中断 1
while(1);                        //等待外部(external interrupt)1 中断
}
```
// ***
//功能:延时 500us 函数 delay500us()
// ***
```c
void delay500us(void)
{
    unsigned char i;
    for(i = 250;i>0;i - -)
    {
        _nop_();
    }
}
```
// ***
//功能:外部中断 1 服务函数(中断编号:2)
// ***
```c
void ExternalInterrupt1(void) interrupt  2   using 1
{
    P3^5 = 0;                    //初始化 P3.5 引脚为低电平
    P3_5 = ~P3_5;                //P3.5 引脚电平取反
//输出 1kHz 的音频信号
//1kHz 的信号周期为 1/1kHz = 1ms,信号电平每 1ms/2 = 500us 取反 1 次
    for(count = 200;count>0;count - -)
    {
        P3_5 = ~P3_5;            //P3.5 引脚电平取反
        delay500us();            //调用延时 500us 函数 delay500us()
    }
//输出 500Hz 的音频信号
//500Hz 信号的周期为 1/500Hz = 2ms,信号电平为每 2ms/2 = 1ms 取反 1 次
    for(count = 200;count>0;count - -)
```

```
{
    P3_5 = ~P3_5;               //对 P3.5 引脚电平取反
    delay500us();               //调用延时 500us 函数 delay500us()
    delay500us();               //调用延时 500us 函数 delay500us()
}
}
```

四、汇编语言源程序

```
;***********************************************************************
;项目名称:报警器设计
;功能:按钮开关 SB 按下时,单片机输出 1kHz 和 500Hz 报警信号
;***********************************************************************
        ORG   0000H
        AJMP  MAIN              ;跳转到主程序
        ORG   0013H             ;外部中断 1 入口地址
        AJMP  INT_1             ;跳转到外部中断 1 服务子程序
;***********************************************************************
;功能:主程序
;***********************************************************************
        ORG   0100H             ;主程序入口地址
MAIN:   MOV   SP,#60H           ;堆栈指针初始化
        SETB  IT1               ;边缘触发方式
        SETB  EA                ;打开中断总开关
        SETB  EX1               ;外部中断 1 允许控制位
        SJMP  $                 ;等待外部 1 中断
;***********************************************************************
;功能:外部中断 1 服务子程序
;***********************************************************************
        ORG   0200H
INT_1:  MOV   P3.5,#00H         ;初始化 P3.5 为低电平,即 P3.5 = 0
START:  MOV   R2,#200           ;设置计数次数 R2 初值为 200
DV1:    CPL   P3.5              ;输出 500Hz 的音频信号
        LCALL DELAY500us        ;调用延时 500us 子程序 DELAY500us
        LCALL DELAY500us        ;调用延时 500us 子程序 DELAY500us
        DJNZ  R2,DV1            ;计数次数 R2 未计满 200 次,跳转到 DV1
        MOV   R2,#200           ;设置计数次数 R2 初值为 200
DV2:    CPL   P3.5              ;输出 1kHz 的音频信号
        LCALL DELAY500us        ;调用延时 500us 子程序 DELAY500us
        DJNZ  R2,DV2            ;计数次数 R2 未计满 200 次,跳转到 DV2
        RETI                    ;中断子程序返回
;***********************************************************************
;功能:延时 500us 子程序 DELAY500us
;***********************************************************************
```

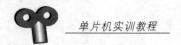

单片机实训教程

```
DELAY500us:MOV R7,#250
LOOP:      NOP
           DJNZ R7,LOOP
           RET
           END
```

实训考核

本课程改革传统的闭卷或开卷考核,而采用过程考核为主的多元化考核方式,考核分为理论考核、职业道德考核和技能考核三部分,各部分所占比例见表17-1、表17-2、表17-3。

表17-1 理论考核和职业素质考核形式及所占比例

序 号	名 称		比 例	得 分
一	理论考核	个人自评	10%	
		组内互评	20%	
		过程作业文件 小组互评	30%	
		老师评定	20%	
		课堂提问、解答	10%	
		项目汇报	10%	
		小计	100%	
二	职业素质	职业道德,工作作风	40%	
		小组沟通协作能力	40%	
		创新能力	20%	
		小计	100%	

表17-2 技能考核内容及比例

姓名		班级		小组		总得分	
序号	考核项目	考核内容及要求		配分	评分标准	考核环节	得分
1	①安全文明生产 ②安全操作规范	着装规范		20%	现场考评	实施	
		安全用电					
		布线规范、合理					
		工具摆放整齐					
		工具及仪器仪表使用规范、摆放整齐					
		任务完成后,进行场地整理,保持场地清洁、有序					

(续表)

2	实训态度	不迟到、早退、旷课	10%	现场考评	六步
		实训过程认真负责			
		组内主动沟通、协作，小组间互助			
3	系统方案制定	工作流程正确合理	10%	现场考评	计划决策
		方案合理			
		选用指令是否合理			
		电路图正确			
	编程能力	独立完成程序	10%	现场考评	决策
		程序简单、可靠			
4	操作能力	正确输入程序并进行程序调试	20%	现场考评	实施
		根据电路图正确接线			
		根据系统功能进行正确操作演示			
5	工艺	接线美观	10%	现场考评	实施
		线路工作可靠			
6	实践效果	系统工作可靠	10%	现场考评	检查
		满足工作要求			
		创新			
		按规定的时间完成项目			
7	汇报总结	工作总结，PPT汇报	5%	现场考评	评估
		填写自我检查表及反馈表			
8	技术文件制作整理	技术文件制作整理能力	5%	现场考评	评估
		合计	100%		

表17-3 各部分考核占课程考核的比例

考核项目	理论考核	技能考核	职业素质考核	合 计
比 例	30%	50%	20%	100%
分 值	30	50	20	100
实际得分				

任务 18 外部计数器中断

实训任务

利用霍尔传感器实现 INT0(P3.2)下降沿中断并计数,在 LCD 上显示出结果,用 K1 按钮清 0,每当用磁铁接近霍尔传感器时,LED 变亮,同时 LCD 上的计数值加 1。

实训设备

1. 设备

PC 机(安装 wave 编程软件、Keil C51 软件)、单片机实验板。

2. 工具及材料

工作对象:电工电子工具、电子元器件和辅助材料、仿真器、编程器。

工作工具:单片机控制电路原理图、实训指导书、项目任务单、工作记录单、项目检查单、各种电工仪表、常用电工工具和拆装工具、量具、相关电子手册。

硬件设计

主控模块采用 ATMEL 公司生产的 AT89S52 单片机。LCD 显示模块选用 1602 字符型 LCD 模块。霍尔传感器模块与单片机的接口电路如图 18-1 所示。

软件设计

一、中断知识

在标准 51 单片机中一般都有两个中断源,分别是 INT0(P3.2)和 INT1(P3.3),其中断入口向量地址分别是 0003H 和 0013H。外部中断共有两种激活方式:一种是电平触发方式,另一种是边沿触发方式。在采用电平触发方式下,在/INTX 引脚上检测到低电平,将触发外部中断,外部中断源应一直保持中断请求有效,直至所请求的中断得到响应为止。在电平触发方式下,当检测到/INTX 引脚出现由高到低的下降沿跳变时将触发外部中断。这两种方式可以通过控制寄存器 TCON 的外部中断方式控制位 IT1 或 IT0 位来控制,当 ITx=1 时,设置为边沿触发方式;当 ITx=0 时,设置为低电平触发方式。对于边沿触发的外部中断,CPU 在响应中断后,就用硬件清除了有关的中断请求标志位 IE0 或 IE1,自动撤除了中断请求。而对于电平触发的外部中断,由于在硬件上 CPU 对/INT0 和/INT1 引脚的信号完全没有控制(在专用的寄存器中,没有相应的中断请求标志),也不像边沿触发方式那样,

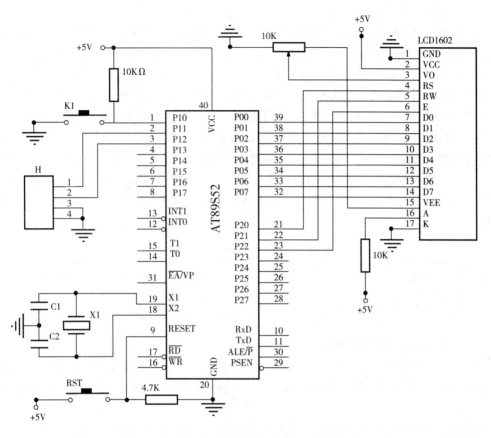

图 18-1 霍尔传感器模块与单片机的接口电路图

响应中断后会自动发出一个应答信号。因此在这种方式下对外部中断源的中断信号有一定的要求：①触发中断的低电平脉冲信号要有一定的宽度；②在中断响应后，触发中断的低电平脉冲信号要能自动跳变为高电平。通常可以利用单稳态触发器对中断源信号进行脉冲整形，使之符合要求。

1. 外部中断的控制与操作

在中断的控制中，与系统外部中断有关的特殊功能寄存器有定时器控制寄存器 TCON（直接地址为 88H）、中断允许寄存器 IE（直接地址为 A8H）和中断优先级管理寄存器 IP（直接地址为 B8H）。

(1) 定时器控制寄存器 TCON

TCON 的位定义如表 18-1 所示。位 TCON.0 和 TCON.2 是外部中断/INT0 和/INT1 的电平触发方式选择位。此位为"0"时，置相应的外部中断为电平触发方式；此位为"1"时，置相应的外部中断为负边沿触发方式。

表 18-1 定时器/计数器控制寄存器 TCON

位地址	8FH	8EH	8DH	8CH	8BH	8AH	89H	88H
位符号	TF1	TR1	TF0	TR0	IE1	IT1	IE0	IT0

(2)中断允许寄存器 IE

IE 主要用于控制中断的开放与禁止,其位功能标识如表 18-2 所示。其中,EA(IE.7):CPU 总中断允许控制。当 EA=1 时,开放 CPU 总中断;EA=0 时,则禁止所有中断。EX0(IE.0):/INT0 中断允许位。当 EX0=1 时,允许外部/INT0 中断;而当 EX0=0 时,外部/INT0 中断被禁止。EX1(IE.2):/INT1 中断允许位。当 EXI=1 时,允许外部/INT1 中断;而当 EX1=0 时,外部/INT1 中断被禁止。ET0(IE.1):定时器/计数器 T0 溢出中断允许位。当 ET0=1 时,允许定时器/计数器 T0 溢出中断;而当 ET0=0 时则禁止定时器/计数器 T0 溢出中断,ET1(IE.3):定时器/计数器 T1 溢出中断允许位。当 ETI=1 时,允许定时器/计数器 T1 溢出中断;而当 ET1=0 时则禁止定时器/计数器 T1 溢出中断。ES(IE.4):串行口中断允许位。当 ES=I 时,允许串行口中断;而当 ES=0 时则禁止串行口中断。

表 18-2 中断允许控制寄存器 IE

位地址	AFH	AEH	ADH	ACH	ABH	AAH	A9H	A8H
位符号	EA	/	/	ES	ET1	EX1	ET0	EX0

(3)中断优先级控制寄存器 IP

IP 主要控制中断响应的优先级。标准 51 的中断分为 2 个优先级。每个中断源的优先级都可以通过中断优先级控制寄存器 IP 中的相应位来设定。表 18-3 给出了 IP 各位的定义,其中:

PX0(IP.0):外部中断 0 优先级设定位。当 PX0=1 时,外部中断 0 被设定为高优先级。
PX1(IP.2):外部中断 1 优先级设定位。当 PX1=1 时,外部中断 1 被设定为高优先级。
PT0(IP.1):定时器 T0 中断优先级设定位。当 PT0=1 时,定时器 T0 被设定为高优先级。
PT1(IP.3):定时器 T1 中断优先级设定位。当 PT1=1 时,定时器 T1 被设定为高优先级。
PS(IP 4):串行口中断优先级设定位。当 PS=1 时,串口中断被设定位高优先级。

表 18-3 中断优先级控制寄存器 IP

位地址	BFH	BEH	BDH	BCH	BBH	BAH	B9H	B8H
位符号	/	/	/	PS	PT1	PX1	PT0	PX0

在默认状态下 5 个中断的优先级关系及相应的入口地址见表 18-4。

表 18-4 MCS-51 单片机的中断入口地址表及中断源自然查询顺序

中断源	优先级	中断入口地址
外部中断/INT0	最高	0003H
定时器 T0 溢出中断		000BH
外部中断/INT1		0013H
定时器溢出中断 T1		001BH
串口发送/接收中断	最低	0023H

任务 18 外部计数器中断

2. 外部中断的应用

(1)实验要求:利用外部中断/INT0 对输入信号进行检测,当每检测到 P3.2 引脚有一个负跳变时计数器加 1。

(2)设计思路:要想使外部中断/INT0 可靠工作,必须对相关的特殊功能寄存器进行初始化设置。设置 EA 位、EX0 位开放总中断和允许/INT0 外部中断,正确设置 IT0 位使外部中断源/INT0 工作于边沿触发方式下。

(3)硬件电路的设计:外部中断/INT0 对应的外部中断源是/INT0 引脚即 P3.2。当外部中断 1 允许且为边沿触发方式时,只要在 P3.2 引脚上出现负的下降沿时,外部中断 0 的标志位 IE0 被置位,CPU 将在下一个机器周期的 S1 状态时响应该中断。实验。

(4)程序设计:主程序主要完成对外部中断进行初始化和数据输出,中断服务子程序主要完成计数。

二、程序流程图

作为练习,请读者画出程序流程图。

三、汇编语言源程序

```
; **************************************************************
;实验名称:外部计数器中断
;功能:利用霍尔传感器实现 INT0 下降沿中断,并计数,在 LCD1602 上显示出结果
; **************************************************************

;LCD 口线的定义
E           BIT     P2.2
RW          BIT     P2.1
RS          BIT     P2.0
LCDPORT     EQU     P0
CMD_BYTE    EQU     2EH
DAT_BYTE    EQU     2FH

KEYBIT      BIT     00H         ;有按键按下标志

DPBL        EQU     30H         ;当前显示的变量(20 到 200 间)
LEDBAI      EQU     31H         ;显示的百位
LEDSHI      EQU     32H         ;显示的十位
LEDGE       EQU     33H         ;显示的个位
LEDSM       EQU     34H         ;现在扫到第几个 LED
KEYTIME     EQU     35H         ;20MS 扫一次按键

            ORG     0000H
            AJMP    MAIN
            ORG     0003H
INT0_1:     INC     DPBL
```

```
            RETI
            ORG     000BH
            AJMP    TIME0_1

;主程序
            ORG     0030H
MAIN:       MOV     20H,#00H
            LCALL   INITLCD
            MOV     DPBL,#00H
            MOV     KEYTIME,#04H
            MOV     CMD_BYTE,#80H       ;设置 DDRAM 的地址
            LCALL   WRITE_CMD
            LCALL   DELAY0
            MOV     DAT_BYTE,#"S"
            LCALL   WRITE_DAT
            MOV     DAT_BYTE,#"U"
            LCALL   WRITE_DAT
            MOV     DAT_BYTE,#"M"
            LCALL   WRITE_DAT
            MOV     DAT_BYTE,#":"
            LCALL   WRITE_DAT
            MOV     TH0,#0EEH           ;16 位的定时器,定时 5MS
            MOV     TL0,#00H
            MOV     TMOD,#01H
            SETB    TR0
            SETB    IT0                 ;INT0 下降沿中断
            MOV     IE,#83H             ;定时器 0 中断
            SJMP    $


            ORG     0100H
TIME0_1:    MOV     TH0,#0EEH
            MOV     TL0,#00H
            DJNZ    KEYTIME,TIME0_RE
            MOV     KEYTIME,#04H        ;每 20MS 进行一次按键扫描
            MOV     P1,#0FFH
            LCALL   KEYSCAN
            LCALL   JSDPBL
            LCALL   DISPDPBL
TIME0_RE:   RETI

;按键扫描
KEYSCAN:    MOV     A,P1
            CJNE    A,#0FEH,CLRKEYBIT
```

```
              JB       KEYBIT,SCAN_RE
              SETB     KEYBIT
              MOV      DPBL,#00H
              AJMP     SCAN_RE
CLRKEYBIT:    CLR      KEYBIT              ;清标志用于等待下一次按键
SCAN_RE:      MOV      P1,#0FFH
              RET

;LCD1602 要用到的一些子程序
;写命令(入口参数 CMD_BYTE)
WRITE_CMD:    CLR      RS
              CLR      RW
              MOV      A,CMD_BYTE
              MOV      LCDPORT,A
              SETB     E
              NOP
              NOP
              CLR      E
              LCALL    DELAY0
              RET

;写显示数据(入口参数 DAT_BYTE)
WRITE_DAT:    SETB     RS
              CLR      RW
              MOV      A,DAT_BYTE
              MOV      LCDPORT,A
              SETB     E
              NOP
              NOP
              CLR      E
              LCALL    DELAY0
              RET

;LCD 显示初始化
INITLCD:      MOV      CMD_BYTE,#30H
              LCALL    WRITE_CMD
              MOV      CMD_BYTE,#30H
              LCALL    WRITE_CMD
              MOV      CMD_BYTE,#30H
              LCALL    WRITE_CMD
              MOV      CMD_BYTE,#38H      ;设定工作方式
              LCALL    WRITE_CMD
              MOV      CMD_BYTE,#0CH      ;显示状态设置
```

```
            LCALL   WRITE_CMD
            MOV     CMD_BYTE,#01H      ;清屏
            LCALL   WRITE_CMD
            MOV     CMD_BYTE,#06H      ;输入方式设置
            LCALL   WRITE_CMD
            RET

;延时子程序
DELAY0:     MOV     R5,#0A0H
DELAY1:     NOP
            DJNZ    R5,DELAY1
            RET

;显示变量
DISPDPBL:   MOV     CMD_BYTE,#85H      ;设置 DDRAM 的地址
            LCALL   WRITE_CMD
            MOV     DAT_BYTE,LEDBAI
            LCALL   WRITE_DAT
            MOV     DAT_BYTE,LEDSHI
            LCALL   WRITE_DAT
            MOV     DAT_BYTE,LEDGE
            LCALL   WRITE_DAT
            RET

;计算显示的值
JSDPBL:     MOV     A,DPBL
            MOV     B,#64H
            DIV     AB
            ORL     A,#30H
            MOV     LEDBAI,A
            MOV     A,B
            MOV     B,#0AH
            DIV     AB
            ORL     A,#30H
            MOV     LEDSHI,A
            MOV     A,B
            ORL     A,#30H
            MOV     LEDGE,A
            RET

            END
```

四、C 语言源程序

```c
// ******************************************************************************
//实验名称:外部中断实验
//功能:利用霍尔传感器实现 INT0 下降沿中断,并计数,在 LCD1602 上显示出结果
// ******************************************************************************
#include <reg51.h>

//LCD 的口线定义
sbit E = P2^2;
sbit RW = P2^1;
sbit RS = P2^0;

unsigned char dpbl,ledbai,ledshi,ledge,keytime;
unsigned char bdata myflag;
sbit keybit = myflag^0;

void time0(void);
void keyscan(void);
void Delay(unsigned int t);
void SendCommandByte(unsigned char ch);
void SendDataByte(unsigned char ch);
void InitLcd(void);
void dispdpbl(void);

void main(void)
{
    InitLcd();
    myflag = 0x00;
    dpbl = 0;
    keytime = 4;              //20MS 扫一次按键
    SendCommandByte(0x80);
    Delay(2);
    SendDataByte('S');
    SendDataByte('U');
    SendDataByte('M');
    SendDataByte(':');
    TH0 = 0xee;               //定时 5MS
    TL0 = 0x00;
    TMOD = 0x01;
    TR0 = 1;
    IT0 = 1;
    IE = 0x83;
```

```c
        while(1)
        {}
}

void int0_1(void)interrupt 0
{
        dpbl++;                 //每进一次下降沿中断,变量计数加 1
}

void time0(void)interrupt 1
{
        TH0 = 0xee;
        TL0 = 0x00;
        if(keytime- - = =0)
        {keytime = 0x04;P1 = 0xff;keyscan();
        ledbai = (dpbl/100)|0x30;
        ledshi = ((dpbl%100)/10)|0x30;
        ledge = (dpbl%10)|0x30;
        dispdpbl();
        }
}

void keyscan(void)
{
        if(P1= = 0xfe)
        {
            if(keybit= =0)   {keybit = 1;dpbl = 0;}
        }
        else                 { keybit = 0;}
        P1 = 0xff;
}

void Delay(unsigned int t)    // delay 40us
{
        for(;t! =0;t- -);
}

// LCD1602 写命令函数
void SendCommandByte(unsigned char ch)
{
        RS = 0;
        RW = 0;
        P0 = ch;
```

```
            E = 1;
                Delay(1);
            E = 0;
            Delay(100);    //delay 40us
    }

// LCD1602 写数据函数
void SendDataByte(unsigned char ch)
{   RS = 1;
        RW = 0;
        P0 = ch;
        E = 1;
            Delay(1);
        E = 0;
        Delay(100); //delay 40us
}

//LCD1602 初始化函数
void InitLcd(void)
{
        SendCommandByte(0x30);
        SendCommandByte(0x30);
        SendCommandByte(0x30);
        SendCommandByte(0x38);       //设置工作方式
        SendCommandByte(0x0c);       //显示状态设置
        SendCommandByte(0x01);       //清屏
        SendCommandByte(0x06);       //输入方式设置
}

void dispdpbl(void)
{
        SendCommandByte(0x85);
        SendDataByte(ledbai);
        SendDataByte(ledshi);
        SendDataByte(ledge);
}
```

实训考核

本课程改革传统的闭卷或开卷考核,而采用过程考核为主的多元化考核方式,考核分为理论考核、职业道德考核和技能考核三部分,各部分所占比例见表 18 - 5、表 18 - 6、表18 - 7。

表 18-5　理论考核和职业素质考核形式及所占比例

序号	名称		比例	得分
一	理论考核	个人自评	10%	
		组内互评	20%	
		过程作业文件 小组互评	30%	
		老师评定	20%	
		课堂提问、解答	10%	
		项目汇报	10%	
		小计	100%	
二	职业素质	职业道德,工作作风	40%	
		小组沟通协作能力	40%	
		创新能力	20%	
		小计	100%	

表 18-6　技能考核内容及比例

姓名		班级		小组		总得分	
序号	考核项目	考核内容及要求		配分	评分标准	考核环节	得分
1	①安全文明生产 ②安全操作规范	着装规范		20%	现场考评	实施	
		安全用电					
		布线规范、合理					
		工具摆放整齐					
		工具及仪器仪表使用规范、摆放整齐					
		任务完成后,进行场地整理,保持场地清洁、有序					
2	实训态度	不迟到、早退、旷课		10%	现场考评	六步	
		实训过程认真负责					
		组内主动沟通、协作,小组间互助					
3	系统方案制定	工作流程正确合理		10%	现场考评	计划决策	
		方案合理					
		选用指令是否合理					
		电路图正确					
	编程能力	独立完成程序		10%	现场考评	决策	
		程序简单、可靠					

(续表)

序号	考核项目	考核内容及要求	配分	评分标准	考核环节	得分
4	操作能力	正确输入程序并进行程序调试	20%	现场考评	实施	
		根据电路图正确接线				
		根据系统功能进行正确操作演示				
5	工艺	接线美观	10%	现场考评	实施	
		线路工作可靠				
6	实践效果	系统工作可靠	10%	现场考评	检查	
		满足工作要求				
		创新				
		按规定的时间完成项目				
7	汇报总结	工作总结,PPT 汇报	5%	现场考评	评估	
		填写自我检查表及反馈表				
8	技术文件制作整理	技术文件制作整理能力	5%	现场考评	评估	
	合计		100%			

表 18-7 各部分考核占课程考核的比例

考核项目	理论考核	技能考核	职业素质考核	合 计
比 例	30%	50%	20%	100%
分 值	30	50	20	100
实际得分				

任务 19 音乐发生器设计

实训任务

设计一个单片机控制的音乐发生器。利用单片机产生"音乐声",从 P3.5 端口输出。

实训设备

1. 设备
PC 机(安装 wave 编程软件、Keil C51 软件)、单片机实验板。
2. 工具及材料
工作对象:电工电子工具、电子元器件和辅助材料、仿真器、编程器。
工作工具:单片机控制电路原理图、实训指导书、项目任务单、工作记录单、项目检查单、各种电工仪表、常用电工工具和拆装工具、量具、相关电子手册。

硬件设计

主控模块采用 ATMEL 公司生产的 AT89S52 单片机,在本项目中选用无源电磁式蜂鸣器来实现报警发声。蜂鸣器的正极接到 V_{cc}(+5V)电源上面,蜂鸣器的负极接到三极管的发射极 E,三极管的基级 B 经过限流电阻 R_1 后由单片机的 P3.5 引脚控制。当 P3.5 输出高电平时,三极管 C8550 截止,没有电流流过线圈,蜂鸣器不发声;当 P3.5 输出低电平时,三极管 C8550 导通,蜂鸣器的电流形成回路,发出声音。因此,可以通过程序控制 P3.5 脚的电平来控制蜂鸣器是否发出声音。音乐发生器与单片机的接口电路如图 19-1 所示。

软件设计

一、算法设计

利用单片机演奏音乐也是单片机爱好者感兴趣的问题之一。这里用实验板来做这个实验,并且了解单片机演奏音乐的基本原理,和相关的源程序。

首先来完成必要的硬件部分,硬件部分比较简单,AT89S52 单片机的 P3.5 口控制一个三极管,三极管控制电磁蜂鸣器的电源通断,电路图见图 19-1。

声音的频谱范围约在几十到几千赫兹,若能利用程序来控制单处机某 I/O 口线的高电平或低电平,则在该口线上就能产生一定频率的矩形波,接上喇叭就能发出一定频率的声音,若再利用延时程序控制"高""低"电平的持续时间,就能改变输出频率,从而改变音调。

任务 19 音乐发生器设计

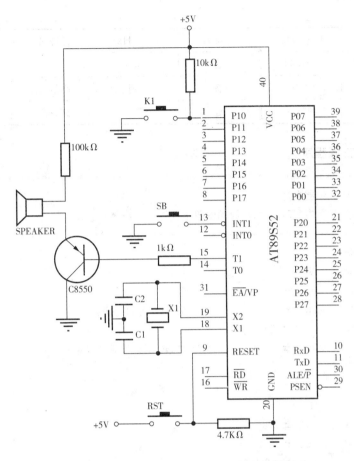

图 19-1 音乐发生器与单片机的接口电路图

例如,要产 200Hz 的音频信号,200Hz 音频的变化周期为 1/200 秒,即 5ms。这样,当 P3.2 的高电平或低电平的持续时间为 2.5ms 时就能发出 200HZ 的音调。在乐曲中,每一音符对应着确定的频率,将每一音符的时间常数和其相应的节拍常数作为一组,按顺序将乐曲中的所有常数排列成一个表,然后由查表程序依次取出,产生音符并控制节奏,就可以实现演奏效果。此外,结束符和休止符可以分别用代码 00H 和 FFH 来表示,若查表结果为 00H,则表示曲子终了;若查表结果为 FFH,则产生相应的停顿效果。

音符与输出频率的对应关系如表 19-1 所示,通过内部定时器,我们可以得到以下的频率。

表 19-1 音符对应的频率表

音符名	频率(Hz)	音符名	频率(Hz)	音符名	频率(Hz)
低音 1	261.63	中音 1	523.25	高音 1	1046.50
低音 2	293.67	中音 2	587.33	高音 2	1174.66
低音 3	329.63	中音 3	659.25	高音 3	1318.51
低音 4	349.23	中音 4	698.46	高音 4	1396.92

(续表)

音符名	频率(Hz)	音符名	频率(Hz)	音符名	频率(Hz)
低音 5	391.44	中音 5	783.99	高音 5	1567.98
低音 6	440	中音 6	880	高音 6	1760
低音 7	493.88	中音 7	987.76	高音 7	1975.52

二、汇编语言源程序

```
;****************************************************************
;实验名称:音乐发生器设计
;功能:利用单片机产生"音乐声",从蜂鸣器放出音乐
;****************************************************************
        ORG    0000H
        LJMP   START
        ORG    000BH
        INC    20H              ;中断服务,中断计数器加 1
        MOV    TH0,#0D8H
        MOV    TL0,#0EFH        ;12M 晶振,形成 10 毫秒中断
        RETI
START:  MOV    SP,#50H
        MOV    TH0,#0D8H
        MOV    TL0,#0EFH
        MOV    TMOD,#01H
        MOV    IE,#82H
MUSIC0: NOP
        MOV    DPTR,#DAT        ;表头地址送 DPTR
        MOV    20H,#00H         ;中断计数器清 0
        MOV    B,#00H           ;表序号清 0
MUSIC1: NOP
        CLR    A
        MOVC   A,@A+DPTR
        JZ     END0             ;是 00H,则结束
        CJNE   A,#0FFH,MUSIC5
        LJMP   MUSIC3
MUSIC5: NOP
        MOV    R6,A
        INC    DPTR
        MOV    A,B
        MOVC   A,@A+DPTR        ;取节拍代码送 R7
        MOV    R7,A
        SETB   TR0              ;启动计数
```

```
MUSIC2:  NOP
         CPL    P3.5
         MOV    A,R6
         MOV    R3,A
         LCALL  DEL
         MOV    A,R7
         CJNE   A,20H,MUSIC2    ;中断计数器(20H)=R7否？；不等,则继续循环
         MOV    20H,#00H        ;等于,则取下一代码
         INC    DPTR
;INC     B
         LJMP   MUSIC1
MUSIC3:  NOP
         CLR    TR0             ;休止100毫秒
         MOV    R2,#0DH
MUSIC4:  NOP
         MOV    R3,#0FFH
         LCALL  DEL
         DJNZ   R2,MUSIC4
         INC    DPTR
         LJMP   MUSIC1
END0:    NOP
         MOV    R2,#64H         ;歌曲结束,延时1秒后继续
MUSIC6:  MOV    R3,#00H
         LCALL  DEL
         DJNZ   R2,MUSIC6
         LJMP   MUSIC0
DEL:     NOP
DEL3:    MOV    R4,#02H
DEL4:    NOP
         DJNZ   R4,DEL4
         NOP
         DJNZ   R3,DEL3
         RET
         NOP

DAT:     DB     18H,   30H,   1CH,   10H
         DB     20H,   40H,   1CH,   10H
         DB     18H,   10H,   20H,   10H
         DB     1CH,   10H,   18H,   40H
         DB     1CH,   20H,   20H,   20H
         DB     1CH,   20H,   18H,   20H
         DB     20H,   80H,   0FFH,  20H
         DB     30H,   1CH,   10H,   18H
```

```
DB    20H,   15H,   20H,   1CH
DB    20H,   20H,   20H,   26H
DB    40H,   20H,   20H,   2BH
DB    20H,   26H,   20H,   20H
DB    20H,   30H,   80H,   0FFH
DB    20H,   20H,   1CH,   10H
DB    18H,   10H,   20H,   20H
DB    26H,   20H,   2BH,   20H
DB    30H,   20H,   2BH,   40H
DB    20H,   20H,   1CH,   10H
DB    18H,   10H,   20H,   20H
DB    26H,   20H,   2BH,   20H
DB    30H,   20H,   2BH,   40H
DB    20H,   30H,   1CH,   10H
DB    18H,   20H,   15H,   20H
DB    1CH,   20H,   20H,   20H
DB    26H,   40H,   20H,   20H
DB    2BH,   20H,   26H,   20H
DB    20H,   20H,   30H,   80H
DB    20H,   30H,   1CH,   10H
DB    20H,   10H,   1CH,   10H
DB    20H,   20H,   26H,   20H
DB    2BH,   20H,   30H,   20H
DB    2BH,   40H,   20H,   15H
DB    1FH,   05H,   20H,   10H
DB    1CH,   10H,   20H,   20H
DB    26H,   20H,   2BH,   20H
DB    30H,   20H,   2BH,   40H
DB    20H,   30H,   1CH,   10H
DB    18H,   20H,   15H,   20H
DB    1CH,   20H,   20H,   20H
DB    26H,   40H,   20H,   20H
DB    2BH,   20H,   26H,   20H
DB    20H,   20H,   30H,   30H
DB    20H,   30H,   1CH,   10H
DB    18H,   40H,   1CH,   20H
DB    20H,   20H,   26H,   40H
DB    13H,   60H,   18H,   20H
DB    15H,   40H,   13H,   40H
DB    18H,   80H,   00H

      END
```

三、C 语言源程序

```c
// *******************************************************************************
//实验名称:音乐发生器设计
//功能:利用单片机产生"音乐声",从蜂鸣器放出音乐
// *******************************************************************************
#include <reg52.h>
#include <ctype.h>
#pragma ot(0)
#define uint   unsigned int
#define uchar unsigned char
#define OSFREQ 11059200           //所使用的晶振频率:11.0592MHZ

/ ************* 音符频率表 *************/
uint code notefreq[] = { 523,587,659,698,784,880,988,1047,1175,1319,1396,1568,1760,1976,
2093,2349,2637,2793,3136,3520,3961};

/ ************* 音名 *************/
uchar code notename[] = {'c','d','e','f',"g",'a','b','1','2','3','4','5','6','7','C','D','E','F','G','A','B',0};

/ ************* 半音频率表 *************  */
uint code halfnotefreq[] = { 554,622,740,831,933,1109,1245,1480,1161,1865,2218,2489,2960,
3322,3729};

/ ************* 音名 *************/
uchar code halfnotename[] = {'c','d','f','g','a','1','2','4','5','6','C','D','F','G','A',0};

sbit   BEEP_PWR = P3^5;
uchar   FreqSandH,FreqSandL;            /* 产生方波的定时器的初值 */
uchar timer1cnt;                         /* 定时器延时计数 */
uchar timer1cntflg;                      /* 定时器定时完成标志 */

/ ********* 定时器 0 用来产生方波 *************/
void timer0int ( )interrupt 1
{
    TH0 = FreqSandH;
    TL0 = FreqSandL;
    BEEP_PWR = ! BEEP_PWR;
}

/ ********** 定时器用来进行比较准确的延时 ************/
void timer1int( )interrupt 3
```

```
        uint * pf;
        if(flg)  { pn = halfnotename; pf = halfnotefreq;}
        else     { pn = notename;     pf = notefreq;}
        while(1)
           {    if(pn[i] = = 0)     return 0;
             if(ch = = pn[i])   return pf[i];
             i + +;
           }
    }

    void Play(char * str)
    {
        uchar i = 0,ch,halfflg = 0;
        uchar lasttime;
        uint freq;
        while(1)
          {for(;;i + +)
             {  ch = str[i];                           //允许曲谱用空格符 '|'符,换行回车等分隔以便阅读
                if((ch = =' ')||(ch = ='|')||(ch = ='\r')||(ch = ='\n')){i + +;
                continue;
             }
             if(! ch){ SoundOff( ); return;}                    //乐曲结束则播放完毕
             if(ch = ='#')   {  halfflg = 1;   continue;}       //半音标志
             if(isdigit(ch)||isalpha(ch))
                { freq = GetFreq(ch,halfflg);                   //从音名获取频率
                     lasttime = 16;
                     break;
                }
             else { halfflg = 0;   continue;}
          }
       i + +;
       ch = str[i];                                             //从下一个符号获取额外音长符号
       while(1)
          {if(! ch)break;
           if(isdigit(ch)||isalpha(ch))break;                   //非音长符号则下次处理
              if(ch = ='-')lasttime + = 8;                      //额外延时一拍
              if(ch = ='.')lasttime + = 4;                      //额外延时半拍
              if(ch = ='_')lasttime/ = 2;                       //下划线相当于简谱中音名下面的下划线,延时减半
    if(ch = ='=')lasttime/ = 4;                                 //双下划线相当于简谱中音名下面的双下划线,延时减为 1/4
           i + +;
           ch = str[i];
          }
     if(freq! = 0)Sound(freq);                                  //发声
```

```
            else       SoundOff();
        delay(lasttime);                              //延时
            }
}

//编谱说明,低音(简谱中数字下面有一个点的)1234567 对应的为小写 cdefgab
//中音(简谱中数字上下都没有点的)1234567 对应的也为 1234567
//高音(简谱中数字上面有一个点的)1234567 对应的为大写 CDEFGAB
//对于降音符 b 或声音符 # 一律用 # + 合适的音名例如 #5
//一个音符本身为一拍,加下划线后为半拍加等号为 1/4 拍
//如:65_4= 则音 6 为一拍,音 5 为半拍,音 4 为 1/4 拍
//下划线或等号连续书写则音长连续变短
//音符后加 - 或 . 表示延长。'-'延长一拍'.'延长半拍多加则延长连续增加


//主函数
void main(void)
{
    //uint i;
    TMOD = 0x11; ET1 = 1; ET0 = 1;   EA = 1;
    while(1)
    {
        // Play("1_1_5_5_6_6_5   4_4_3_3_2_2_1   5_5_4_4_3_3_2 5_5_4_4_3_3_21_1_5_5_6_6_5   4_4_3_3_2_2_1"); //满天都是小星星
        Play("1_2_3_1_   1_2_3_1_   3_4_5 3_4_5   5=6=5=4=3_1_ 5=6=5=4=3_1_ 2_g_12_g_1");
        //两只老虎
        //Play("a-a1-a2--a-b1b13-2a--a--   a-33-12--a-b1b13-21--1--5-55432--a-b1-12123--3--1-1_1_1235--4-32-b3-2a--a--   a-66565--4-34-56543--3--1-1_1_1235--4-32-b3-2a--a--"); //山楂树
        Play("5._3=2_1_5-12_3_g-5.3_23_5_1a_3_2-356.5_352._3=2_1_a32_21_a1g05.3_6562_3_50"); //学习雷锋好榜样
        //Play("C-53.2_1530C-53.2_1650 5_C6_5_C05_C6_5_6_0_3_C.6_53C.6_C0C53_6_5_3_2.1_30_5_C56_C_6_5_33_1_6-60C._C=5_5_2._3=5_5_6.5_6DC6_5_C6_5_33_5_C-");
        // Play("3-2_3_4_33-2_3_4_33-4-3_4_5_4 4-3-2-  3-2_3_4_33-2_3_4_33-4-3_4_5_4 4-3-2"); //许巍-星空前奏
        Play("543 3_2_1_2_30 g53 3_2_1_2_30 a65 4_3_2_3_4 1 b a g");      //许巍-时光
        SoundOff( );
            }
}
```

实训考核

本课程改革传统的闭卷或开卷考核,而采用过程考核为主的多元化考核方式,考核分为理论考核、职业道德考核和技能考核三部分,各部分所占比例见表 19-2、表 19-3、表 19-4。

表 19-2 理论考核和职业素质考核形式及所占比例

序号	名称		比例	得分
一	理论考核	个人自评	10%	
		组内互评	20%	
		过程作业文件 小组互评	30%	
		老师评定	20%	
		课堂提问、解答	10%	
		项目汇报	10%	
		小计	100%	
二	职业素质	职业道德,工作作风	40%	
		小组沟通协作能力	40%	
		创新能力	20%	
		小计	100%	

表 19-3 技能考核内容及比例

姓名		班级		小组		总得分	
序号	考核项目	考核内容及要求		配分	评分标准	考核环节	得分
1	①安全文明生产 ②安全操作规范	着装规范		20%	现场考评	实施	
		安全用电					
		布线规范、合理					
		工具摆放整齐					
		工具及仪器仪表使用规范、摆放整齐					
		任务完成后,进行场地整理,保持场地清洁、有序					
2	实训态度	不迟到、早退、旷课		10%	现场考评	六步	
		实训过程认真负责					
		组内主动沟通、协作,小组间互助					
3	系统方案制定	工作流程正确合理		10%	现场考评	计划决策	
		方案合理					
		选用指令是否合理					
		电路图正确					
	编程能力	独立完成程序		10%	现场考评	决策	
		程序简单、可靠					

（续表）

序号	考核项目	考核内容及要求	配分	评分标准	考核环节	得分
4	操作能力	正确输入程序并进行程序调试	20%	现场考评	实施	
		根据电路图正确接线				
		根据系统功能进行正确操作演示				
5	工艺	接线美观	10%	现场考评	实施	
		线路工作可靠				
6	实践效果	系统工作可靠	10%	现场考评	检查	
		满足工作要求				
		创新				
		按规定的时间完成项目				
7	汇报总结	工作总结，PPT汇报	5%	现场考评	评估	
		填写自我检查表及反馈表				
8	技术文件制作整理	技术文件制作整理能力	5%	现场考评	评估	
		合计	100%			

表19-4 各部分考核占课程考核的比例

考核项目	理论考核	技能考核	职业素质考核	合计
比例	30%	50%	20%	100%
分值	30	50	20	100
实际得分				

任务 20 99.9秒马表设计

实训任务

设计一个单片机控制的马表。产生秒计数器,使马表从 00.0 计时到 99.9。当计时到 99.9 后,系统计时溢出,重新回零 00.0 计数,周而复始。

实训设备

1. 设备

PC 机(安装 wave 编程软件、Keil C51 软件)、单片机实验板。

2. 工具及材料

工作对象:电工电子工具、电子元器件和辅助材料、仿真器、编程器。

工作工具:单片机控制电路原理图、实训指导书、项目任务单、工作记录单、项目检查单、各种电工仪表、常用电工工具和拆装工具、量具、相关电子手册。

硬件设计

主控模块采用 ATMEL 公司生产的 AT89S52 单片机。

秒表显示模块由 3 个共阳极数码管构成,分别用来显示时间的十位计数单元值、个位计数单元值和 0.1s 计数单元值。由于 LED 的位数比较多,数码管采用静态显示方式时,要占用大量的 I/O 口线,硬件电路比较复杂。为了简化电路,降低成本,在本项目中数码管采用动态显示方式。数码管的公共使能端 COM 连接三极管 C8550 的集电极,三极管 C8550 主要用于信号的放大,以驱动数码管工作。3 个三极管 C8550 的基极通过限流电阻分别接到单片机 P2 口的 P2.3、P2.4、P2.5,通过控制三极管 C8550 的基极电平来打开或关闭数码管的显示,起到"使能"作用。三极管 C8550 的发射极接+5V 电源,数码管显示模块与单片机的接口电路如图 20-1 所示。

软件设计

一、算法设计

秒表设计需要采用比较精确的计时,可以利用 MCS-51 单片机内部的可编程定时器/计数器来实现精确的计时。

1. 程序实现方式

单片机的定时/计数可以采用查询方式或者中断方式来实现,本项目采用中断方式。

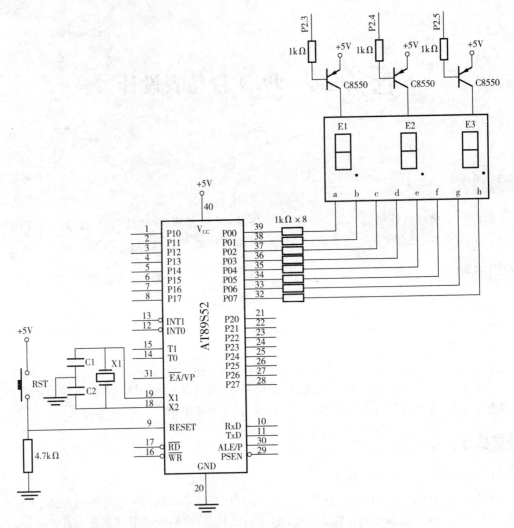

图 20-1 马表显示模块与单片机的接口电路原理图

2. 确定定时器的工作方式

由于秒表的最小计数单位为 0.1s,而单片机 16 位定时器的最大定时时间分别为(假设单片机系统的晶振为 12MHz):

方式 0: $t_{max} = 2^{13} \times 12/f_{OSC} = 8192 \times 12/(12 \times 10^6) = 8.192(ms)$;

方式 1: $t_{max} = 2^{16} \times 12/f_{OSC} = 65536 \times 12/(12 \times 10^6) = 65.536(ms)$;

方式 2: $t_{max} = 2^8 \times 12/f_{OSC} = 256 \times 12/(12 \times 10^6) = 0.256(ms)$。

可见,上述三种方式下的最大定时时间都不能达到秒表的最小计数单位 0.1s,在这种状况下,为了减少中断或定时到的次数,避免响应误差或中间重置误差,使定时更精确,选用定时时间最长的方式,即方式 1。

3. 确定基本定时时间

确定基本定时时间的原则是:基本定时时间尽量长且必须与要求的定时时间成整数倍关系。据此可选择定时器的基本定时时间为 5ms,控制软计数器的累计次数为 20 次,即可实现 0.1s(5ms×20＝100ms＝0.1s)的定时要求。

4. 定时初值计算

定时器 T0 定时基本时间为 5ms,单片机系统所用的石英晶体振荡频率为 11.0592MHz。因此,1 个机器周期＝1/石英频率×12,即为 12/11.0592us,定时器工作方式设置为方式 1,计算初值如下:

$$X=2^{16}-t\times f_{osc}/12=65536-5\times10^{-3}\times11.0592\times10^{6}/12=60928=0xee00$$

所以 TH0＝0xee,TL0＝0x00

5. 系统工作原理

当从 00.0 计时到 0.1s 后,0.1s 计数单元由 0 变为 1,再计时到 0.1s,0.1s 计数单元由 1 变为 2,以此类推……

当 0.1s 计数单元变为 9 后,如果再计时到 0.1s,则个位计数单元由 0 变为 1;再计时到 1s,个位计数单元由 1 变为 2,以此类推……

当个位计数单元变为 9 后,如果再计时到 1s,则十位计数单元由 0 变为 1;再计时到 10s,十位计数单元由 1 变为 2,以此类推……

当计数到 99.9 后,系统计时溢出,重新回零计数,周而复始。

二、程序设计

1. 主函数设计

主函数主要完成对定时器和有关寄存器及变量进行初始化,然后等待定时器 T0 中断的到来。

(1)初始化

计数器小数位 js100ms 清零,计数器个位 jsgw 清零,计数器十位 jssw 清零,系统上电时从 00.0 开始显示。设置软计数器 keytime 初值为 0x14,即 20 次。设置数码管号数 lednumber 初值为 0x01。

(2)定时器设置

设定定时器 T0 的工作方式为方式 1,即寄存器 TMOD＝0x01。启动定时器 T0,即 TR0＝1。开放总中断及定时器 T0 中断,即设定 IE＝0x82。

(3)等待中断

定时器 T0 启动计时后,CPU 等待定时中断的到来。当定时器 T0 定时 5ms 后,进入定时器 T0 中断服务函数。

主函数设计流程图如图 20-2 所示。

2. 定时器 T0 中断服务函数设计

定时器 T0 中断服务函数设计分为计时部分和显示部分。

当定时器 T0 定时 5ms 后,进入定时器 T0 中断服务函数,重装定时器 T0 初值,即 TH0＝0xee,TL0＝0x00。每定时 5ms 一次,软计数器 keytime 值减 1。

(1)计时部分设计如下:

① 首先判断软计数器 keytime 是否到 0(即 0.1s)。

a. 若 keytime 为 0,表明 0.1s 计时到,重置软计数器 keytime 初值为 20,然后将 0.1s 计数单元 js100ms 加 1。

b. 若 keytime 不为 0,表明 0.1s 计时未到,则程序转向 LED 显示部分。

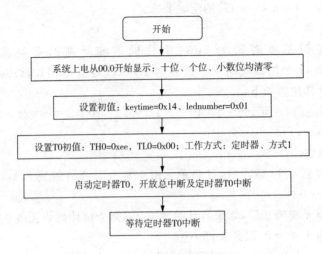

图 20-2 主函数设计流程图

② 再判断 0.1s 计数单元 js100ms 是否到 10(即 1s)。

a. 若 js100ms 为 10,则将 0.1s 计数单元 js100ms 清零,然后将秒个位计数单元 jsgw 加 1。

b. 若 js100ms 未到 10,则程序转向 LED 显示部分。

③ 接着判断秒个位计数单元 jsgw 是否到 10(即 10s)。

a. 若 jsgw 为 10,则将秒个位计数单元 jsgw 清零,然后将秒十位计数单元 jssw 加 1。

b. 若 jsgw 未到 10,则程序转向 LED 显示部分。

④ 最后判断秒十位计数单元 jssw 是否到 10(即 100s)。

a. 若 jssw 为 10,则系统计数溢出,将秒十位计数单元 jssw 清零,系统重新回零计数。

b. 若 jssw 未到 10,则程序转向 LED 显示部分。

(2) 显示部分设计如下:

关显示,数码管号数 lednumber+1→lednumber。

① 首先判断 lednumber 是否为 4。

a. 若 lednumber 为 4,重置 lednumber 为 1,程序转向判断 lednumber 是否为 1 或 2 或 3。

b. 若 lednumber 不为 4,程序转向判断 lednumber 是否为 1 或 2 或 3。

② 然后判断 lednumber 是否为 1 或 2 或 3。

a. 若 lednumber 为 1,驱动第一个数码管,即 P2=0xdf,并把 ledcode[js100ms]送给 P0 显示后程序退出。

b. 若 lednumber 为 2,驱动第二个数码管,即 P2=0xef,并把 ledcode[jsgw]&0xfb 送给 P0 显示后程序退出。

c. 若 lednumber 为 3,驱动第三个数码管,即 P2=0xf7,并把 ledcode[jssw]送给 P0 显示后程序退出。

d. 若 lednumber 是否为 1 或 2 或 3,程序退出。

③ 定时器 T0 中断服务函数设计流程图如图 20-3 所示。

任务20 99.9秒马表设计

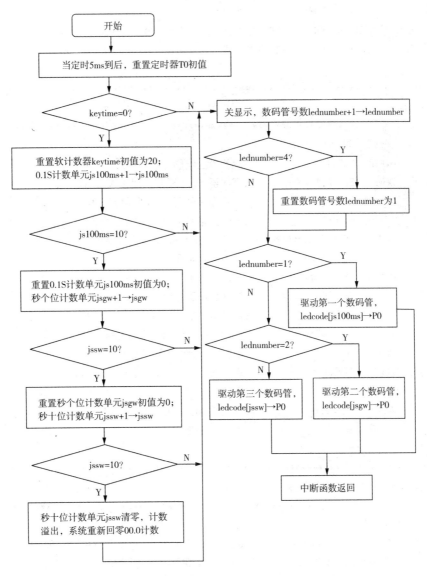

图 20-3 定时器 T0 中断服务函数设计流程图

三、汇编语言源程序

```
;**********************************************************************
;项目名称:单片机控制的马表设计
;功能:利用定时器,产生秒计数器,使数码管从00.0计到99.9
;**********************************************************************


;**********************************************************************
;功能:数据结构定义
;**********************************************************************
JS100MS     EQU     30H     ;0.1S计数单元
JSGW        EQU     31H     ;个位计数单元
```

```
JSSW            EQU     32H             ;十位计数单元
KEYTIME         EQU     33H             ;软计数器,产生 0.1S = 5ms * 20 = 100ms
LEDSM           EQU     34H             ;数码管号数

;****************************************************************************
;功能:主程序
;****************************************************************************
                ORG     0000H
                AJMP    MAIN                    ;跳转到主程序
                ORG     000BH                   ;定时器 T0 中断服务程序入口地址
                AJMP    TIME0_1                 ;跳转到定时器 T0 中断服务子程序
MAIN:           ORG     0050H                   ;系统上电时从 00.0 开始显示
                MOV     JS100MS,#00H            ;0.1S 计数单元 JS100MS 清零
                MOV     JSGW,#00H               ;个位计数单元 JSGW 清零
                MOV     JSSW,#00H               ;十位计数单元 JSSW 清零
                MOV     KEYTIME,#14H            ;软计数器置初值,5ms×20 = 100ms = 0.1s
                MOV     LEDSM,#01H              ;扫描第一个数码管
                MOV     TH0,#0EEH               ;定时 5ms,定时器 T0 置初值:TH0 = 0EEH
                MOV     TL0,#00H                ;定时 5ms,定时器 T0 置初值:TL0 = 00H
                MOV     TMOD,#01H               ;定时器 T0 定时方式设置:定时、方式 1
                SETB    TR0                     ;启动定时器 T0
                MOV     IE,#82H                 ;开放总中断与定时器 T0 中断
                SJMP    $                       ;等待 T0 中断

;****************************************************************************
;功能:定时器 T0 中断服务子程序
;****************************************************************************
                ORG     0100H
TIME0_1:        MOV     TH0,#0EEH               ;定时 5ms,定时器 T0 重置初值:TH0 = 0EEH
                MOV     TL0,#00H                ;定时 5ms,定时器 T0 置初值:TL0 = 00H
                DJNZ    KEYTIME,TODISP          ;判断软计数器 KEYTIME 减 1 后是否为 0
                MOV     KEYTIME,#14H            ;软计数器重置初值
                INC     JS100MS                 ;0.1S 计数单元 JS100MS 加 1
                MOV     A,JS100MS               ;0.1S 计数单元 JS100MS 送给累加器
                CJNE    A,#0AH,TODISP           ;判断 0.1S 计数单元 JS100MS 是否计满 10 次
                                                ;0.1S 计数单元计满 10 次,开始计数个位
                MOV     JS100MS,#00H            ;重新初始化 0.1S 计数单元
                INC     JSGW                    ;个位计数单元 JSGW 加 1
                MOV     A,JSGW                  ;个位计数单元 JSGW 送给累加器
                CJNE    A,#0AH,TODISP           ;个位计数单元 JSGW 计满 10 次,开始计数十位
                MOV     JSGW,#00H               ;重新初始化个位计数单元 JSGW
                INC     JSSW                    ;十位计数单元 JSSW 加 1
                MOV     A,JSSW                  ;十位计数单元 JSSW W 送给累加器
```

任务 20　99.9 秒马表设计

```
            CJNE    A,#0AH,TODISP        ;十位计数单元 JSSW 计满 10 次,重新回零计数
            MOV     JSSW,#00H            ;溢出 99.9,则回到 00.0

TODISP:     MOV     P0,#0FFH             ;在位选时关显示
            INC     LEDSM                ;数码管号数 LEDSM 加 1
            MOV     A,LEDSM              ;数码管号数 LEDSM 送给累加器
            CJNE    A,#04H,TODISP1       ;判断数码管号数 LEDSM 是否为 4
            MOV     LEDSM,#01H           ;扫完第三个又从第一个开始

TODISP1:    MOV     A,LEDSM
            CJNE    A,#01H,TODISP2       ;判断数码管号数 LEDSM 是否为 1
            MOV     P2,#0DFH             ;P2.5=0,驱动第一个数码管
            MOV     A,JS100MS            ;0.1S 计数单元送给累加器
            MOV     DPTR,#LEDCODE        ;表头地址送给 DPTR
            MOVC    A,@A+DPTR            ;查表
            AJMP    TOP0                 ;跳转到数码管显示程序

TODISP2:    CJNE    A,#02H,TODISP3       ;判断数码管号数 LEDSM 是否为 2
            MOV     P2,#0EFH             ;P2.4=0,驱动第二个数码管
            MOV     A,JSGW               ;个位计数单元 JSGW 值送给累加器
            MOV     DPTR,#LEDCODE        ;表头地址送给 DPTR
            MOVC    A,@A+DPTR            ;查表
            CLR     ACC.2                ;把 H 点(第二个数码管小数点)点亮
            AJMP    TOP0                 ;跳转到数码管显示程序

TODISP3:    MOV     P2,#0F7H             ;P2.3=0,驱动第三个数码管
            MOV     A,JSSW               ;十位计数单元 JSSW 值送给累加器
            MOV     DPTR,#LEDCODE        ;表头地址送给 DPTR
            MOVC    A,@A+DPTR            ;查表

TOP0:       MOV     P0,A                 ;查表取得显示编码送到数码管显示单元

            RETI                         ;定时器 T0 中断服务子程序返回

;************************************************************
;功能:LED 字段码表
;************************************************************
LEDCODE:DB   0C0H,0F9H,0A4H,0B0H,99H,92H,82H,0F8H,80H,90H

            END
```

四、C 语言源程序

//**

```c
//实验名称:秒表设计
//功能:利用定时器,产生秒计数器,使数码管从0.00计到99.9
//*****************************************************************************
#include <reg51.h>              //包含"51寄存器定义"头文件


//*****************************************************************************
//功能:变量定义
//*****************************************************************************
//*****************************************************************************
//程序中不变数据存放在片内的CODE存储区,以节省宝贵的RAM
//ledcode[ ]—存储数码管字形码
//*****************************************************************************
unsigned char codeledcode[ ]={0xc0,0xf9,0xa4,0xb0,0x99,0x92,0x82,0xf8,0x80,0x90};
//js100ms—小数位计数单元
//jsgw—个位计数单元
//jssw—十位计数单元
//keytime—软计数器(用于产生0.1s)
//lednumber—数码管号数


unsigned char js100ms,jsgw,jssw,keytime,lednumber;
//*****************************************************************************
//功能:函数声明
//*****************************************************************************
void time0(void);                //定时器T0中断服务函数


//*****************************************************************************
//功能:主函数
//*****************************************************************************
void main(void)                  //主函数入口地址
{
    js100ms=0x00;                //马表计数器小数位清零
    jsgw=0x00;                   //马表计数器个位清零
    jssw=0x00;                   //马表计数器十位清零
    keytime=0x14;                //软计数器置初值,5ms×20=100ms=0.1s
    lednumber=0x01;              //扫描第一个数码管,从第一个LED开始显示
    TH0=0xee;                    //定时5ms,定时器TH0置初值
    TL0=0x00;                    //定时5ms,定时器TL0置初值
    TMOD=0x01;                   //定时器T0定时方式设置:定时器,工作方式1
    TR0=1;                       //启动定时器T0
    IE=0x82;                     //打开定时器T0中断
    while(1)                     //等待定时器T0中断
    {   }
}
```

// **
//功能:定时器 T0 中断服务函数
// **
void time0(void)interrupt 1
{
        //计时部分程序代码
        TH0 = 0xee;   //定时 5ms,重新初始化定时器 TH0 置初值
        TL0 = 0x00;   //定时 5ms,重新初始化定时器 TL0 置初值
        keytime - - ;   //每定时 1ms 1 次,软计数器 keytime 自减 1
        if(keytime = = 0x00)   //判断软计数器 keytime 值是否为 0
        {
          keytime = 0x14;   //软计数器 keytime 重置初值 20
          js100ms + + ;   //马表计数器小数位加 1
          if(js100ms = = 0x0a)   //判断马表小数位是否计满 10 次
          {
            js100ms = 0x00;   //马表计数器小数位清零
            jsgw + + ;   //马表计数器个位加 1
            if(jsgw = = 0x0a)   //判断马表个位是否计满 10 次
            {
              jsgw = 0x00;   //马表计数器个位清零
              jssw + + ;   //马表计数器十位加 1
              if(jssw = = 0x0a)   //判断马表十位是否计满 10 次
              { jssw = 0x00;}   //马表计数器十位清零
            }
          }
        }

        //显示部分程序代码
        lednumber + + ;   //数码管号数增 1
        if(lednumber = = 0x04)   //判断数码管号数是否为 4
        { lednumber = 0x01;}   //若数码管号数为 4,重新从第一数码管开始显示
        P0 = 0xff;   //在数码管位选时关掉显示
        switch(lednumber)   //判断数码管号数 lednumber 的值
        {
          //数码管号数 lednumber 为 1,则驱动数码管小数位计数单元片选信号 P2.5 = 0
          //同时把控制数码管显示字形值 ledcode[s100ms]送给 P0,退出中断
          case 0x01: { P2 = 0xdf;P0 = ledcode[js100ms];}   break;

          //数码管号数 lednumber 为 2,则驱动数码管个位计数单元片选信号 P2.4 = 0
          //同时把控制数码管显示字形值 ledcode[jsgw]送给 P0,并点亮小数点
          case 0x02: { P2 = 0xef;P0 = ledcode[jsgw]&0xfb;}   break;
          //数码管号数 lednumber 为 3,则驱动数码管十位计数单元片选信号 P2.3 = 0

```
            //同时把控制数码管显示字形值 ledcode[jssw]送给 P0,退出中断
            case 0x03: { P2 = 0xf7;P0 = ledcode[jssw];}        break;

            //缺省值时退出中断服务程序
            default: break;
        }
    }
```

实训考核

本课程改革传统的闭卷或开卷考核,而采用过程考核为主的多元化考核方式,考核分为理论考核、职业道德考核和技能考核三部分,各部分所占比例见表 20-1、表 20-2、表 20-3。

表 20-1 理论考核和职业素质考核形式及所占比例

序号	名称			比例	得分
一	理论考核	过程作业文件	个人自评	10%	
			组内互评	20%	
			小组互评	30%	
			老师评定	20%	
		课堂提问、解答		10%	
		项目汇报		10%	
		小计		100%	
二	职业素质	职业道德,工作作风		40%	
		小组沟通协作能力		40%	
		创新能力		20%	
		小计		100%	

表 20-2 技能考核内容及比例

姓名		班级		小组		总得分	
序号	考核项目	考核内容及要求		配分	评分标准	考核环节	得分
1	①安全文明生产 ②安全操作规范	着装规范		20%	现场考评	实施	
		安全用电					
		布线规范、合理					
		工具摆放整齐					
		工具及仪器仪表使用规范、摆放整齐					
		任务完成后,进行场地整理,保持场地清洁、有序					

（续表）

序号	考核项目	考核内容及要求	配分	评分标准	考核环节	得分
2	实训态度	不迟到、早退、旷课	10%	现场考评	六步	
		实训过程认真负责				
		组内主动沟通、协作、小组间互助				
3	系统方案制定	工作流程正确合理	10%	现场考评	计划决策	
		方案合理				
		选用指令是否合理				
		电路图正确				
	编程能力	独立完成程序	10%	现场考评	决策	
		程序简单、可靠				
4	操作能力	正确输入程序并进行程序调试	20%	现场考评	实施	
		根据电路图正确接线				
		根据系统功能进行正确操作演示				
5	工艺	接线美观	10%	现场考评	实施	
		线路工作可靠				
6	实践效果	系统工作可靠	10%	现场考评	检查	
		满足工作要求				
		创新				
		按规定的时间完成项目				
7	汇报总结	工作总结，PPT汇报	5%	现场考评	评估	
		填写自我检查表及反馈表				
8	技术文件制作整理	技术文件制作整理能力	5%	现场考评	评估	
	合计		100%			

表 20-3 各部分考核占课程考核的比例

考核项目	理论考核	技能考核	职业素质考核	合　计
比　例	30%	50%	20%	100%
分　值	30	50	20	100
实际得分				

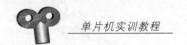

任务 21 RS-232 串行通信

实训任务

设计一个 PC 机与单片机串行通信的控制系统。单片机系统接收 PC 机串口发送过来的 ASCII 码,并在 LCD 上显示。同时单片机把接收到的 ASCII 码回传给 PC 机串口并显示,以验证串口接收、发送数据的正确性。

实训设备

1. 设备

PC 机(安装 wave 编程软件、Keil C51 软件)、单片机实验板。

2. 工具及材料

工作对象:电工电子工具、电子元器件和辅助材料、仿真器、编程器。

工作工具:单片机控制电路原理图、实训指导书、项目任务单、工作记录单、项目检查单、各种电工仪表、常用电工工具和拆装工具、量具、相关电子手册。

硬件设计

主控模块采用 ATMEL 公司生产的 AT89S52 单片机。LCD 显示模块选用 1602 字符型 LCD 模块。

使用 RS-232 接口进行异步通信,必须将单片机的 TTL 电平转换为 RS-232 电平,即在通信方的单片机接口部分增加 RS-232 电气转换接口,在本项目中利用 MAXIM 公司的 MAX232 集成芯片构成转换接口电路。串行通信模块与单片机的接口电路原理图如图 21-1 所示。

软件设计

在实际应用中,单片机往往不是作为一个独立的控制单元而存在的,它还要和其他控制单元进行通信,这些控制单元可以是另一个单片机,也可以是 PC 机。单片机与单片机之间的通信相对比较简单。本节着重讨论的是如何实现单片机与单片机之间的串行通信传输,和基于 MAX232 的单片机与 PC 之间的通信。

一、串行通信基础

1. 同步通信与异步通信

串行通信中的数据线上的数据是具有瞬时性的,某位数据只能存在于某一特定的时间

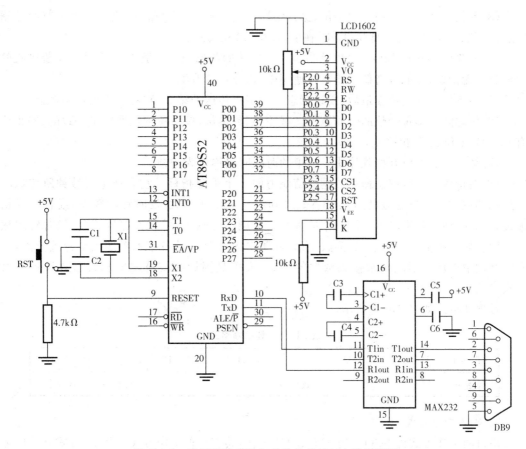

图 21-1 串行通信模块与单片机的接口电路原理图

段内,如果接收方在该时间段内没有对数据线进行读取,则该位数据即会丢失。这就要求当一方在发送数据时,另一方必须也在同步接收。同步是指接收方在数据线上读得某位数据的频率与发送方在数据线上发送某位数据的频率必须是严格一致的。但采用同步通信的方式进行通信时,很难保证收发双方的时钟严格一致。如果收发双方的时钟有微小的偏差,即使不至于影响单个字符的接收,但在传送批量数据时出现的误差累积效应,达到一定程度时足以使接收数据出错。而且如果接收方由于某种原因(如噪声等)漏掉一位,则所有以下接收的数据都是不正确的。一般人们利用异步通信方式来解决同步通信方式的不足。

异步通信是指发送和接收双方分别使用自己的时钟,以单个字符作为最小发送/接收单元。也就是说,在这种通信方式下,每个字符作为独立的信息单元,可以随机地出现在数据流中,而每个字符出现在数据流中的相对时间是随机的,由此可见,所谓的"异步"是指字符与字符间的异步,而在每个字符内部,收发双方的时钟频率仍然需要保持同步。这有效解决了同步通信方式的弊端。在本开发板上,有 1 个可以和单片机通信 DB9 针接口。

2. 异步通信的基本概念

(1)在信号线上共有两种状态,可分别用逻辑 1 和逻辑 0 来区分。在发送器空闲时,数据线应该保持在逻辑 1 状态。

(2)起始位(Start Bit)。发送器是通过发送起始位而开始一个字符传送,起始位使数据线处于逻辑 0 状态,提示接收器数据传输即将开始。

(3)数据位(Data Bits)。起始位之后就是传送数据位。数据位一般为一个字节的数据(也有 6 位、7 位的情况),低位(LSB)在前,高位(MSB)在后。

(4)校验位(parity Bit)。可以认为是一个特殊的数据位。校验位一般用来判断接收的数据位有无错误,一般是奇偶校验。在使用中,该位常常取消。

(5)停止位。停止位在最后,用以标志一个字符传送的结束,它对应于逻辑 1 状态。

(6)位时间。即每个位的时间宽度。起始位、数据位、校验位的位宽度是一致的,停止位有 0.5 位、1 位、1.5 位格式,一般为 1 位。

(7)帧。从起始位开始到停止位结束的时间间隔称之为一帧。

(8)波特率。波特率是衡量数据通信能力的一个重要指标。它表征了每秒钟所传送的二进制的位数。单位是波特/秒(bit/s)。同时,它也反映了收/发时钟的频率,或者更确切地说,收方时钟的频率决定了波特率的大小。在同步通信中,波特率等同于收/发时钟的频率。而在异步通信中,其值一般为收/发时钟频率的 1/16。波特率的设定是十分重要的,在实际开发中出现的不能正常接收数据的现象,大多是因影响波特率的相关参数设置不对所致。

数据帧格式如表 21-1 所示。

表 21-1 数据帧格式

START	D0	D1	D2	D3	D4	D5	D6	D7	P	STOP
起始位	数据位								校验位	停止位

3. 串口工作方式

串口的工作方式由 SCON 的 SM0、SMI 定义,共有 4 种工作方式。其中,方式 0 是作为同步移位寄存器,可以通过外接移位寄存器芯片实现扩展 I/O 接口功能,一般不用于通信,故在此不作介绍。其他 3 种方式均是异步通信方式。

方式 1。8 位数据异步通信接口。波特率可变。波特率由定时器 T1 或 T2 的溢出频率经分频后得到。

方式 2。9 位数据异步通信接口。波特率由主频分频得到,当 SMOD=1 时,波特率为 fosc/32;当 SMOD=0 时,波特率为 fosc/64。

方式 3。9 位数据异步通信接口。波特率可变。波特率由定时器 Tl 或 T2 的溢出频率经分频后得到。

4. 波特率的选择与设置

串行口不同的工作方式,所需的波特率发生器亦不同。方式 1、方式 3 的波特率发生器可由定时器 T1、T2 提供。而方式 2 的波特率是固定的,只能由主频分频后得到。当我们选择定时器作为波特率发生器时,通常设置定时器 1 工作在方式 2,由于定时器 1 的方式 2 为自动重装入 8 位计数方式,无需中断服务程序,只需对其进行初始化。此时,定时器 TI 的溢出率与波特率成正比。定时器 1 的溢出率可通过公式 4-1 求得

$$f_{溢出} = (fosc/12) \times (1/(256-TH1)) \qquad (4-1)$$

特殊功能寄存器 PCON 中的 SMOD 位为串行口波特率控制位,当 SMOD=1 时,使波特率加倍。故波特率的计算可通过公式 4-2 获得:

$$f_{bit} = (2^{SMOD}/32) \times (f_{osc}/12) \times (K/(256-TH1)) \qquad (4-2)$$

式中：f_{osc}——晶振频率；

f_{bit}——所要设置的波特率；

SMOD——对应 PCON 中的 SMOD 位；

K——分频系数。

5. 单片机收/发数据流程

MCS—51 单片机串行口发送/接收数据时，通过两个串行缓冲器 SBUF 进行，这两个缓冲器采用一个地址（98H），但在物理上是独立的。其中接收缓冲器只能读出不能写入，发送缓冲器只能写入不能读出。

发送过程。由指令 MOV SBUF,A 启动,此时待传送的数据由 A 累加器传入串行发送缓冲器 SBUF,由硬件自动在发送字符的始、末加上起始位（低电平）、停止位（高电子）及其他控制位（如奇偶位等），而后在移位脉冲的控制下,低位在前,高位在后,逐位从 TXD 端（方式 0 除外）发出。

接收过程。串行口的接收与否受制于允许接收位 REN 的状态,当 REN 被软件置"1"后,允许接收器接收。串口的接收器以所选波特率的 16 倍速对 RXD 线进行监视。当"1"到"0"跳变时,检测器连续采样到 RXD 线上低电平时。便认定 RXD 端出现起始位,继而接收控制器开始工作。在每位传送时间的第 7、8、9 三个脉冲状态采样 RXD 线,决定所接收的值为"0"或"1"。当接收完停止位后,控制电路使中断标志 R1 置为"1",此时程序可通过 MOV A,SBUF 指令将接收到的字符从 SBUF 送入累加器 A,从而完成一帧数据的接收工作。

二、编写单片机异步通信程序

1. 单片机与单片机的通信程序

单片机异步通信程序,依接收数据方式的不同而分为查询方式和中断方式。查询方式由其固有的缺点只能应用于小数据量的传送,一般不推荐采用此种方式。采用中断接收方式时需要对中断允许寄存器 IE 的进行设置。该寄存器的 2 位与串行口中断有关：MCU 中断允许位 EA,当 EA=1 时,MCU 允许中断,EA=0 时,MCU 屏蔽一切中断请求；串行口中断允许位 ES,当 ES=1 时,允许串行口中断,ES=0 时,禁止串行接口申请中断。所以在采用中断方式编程时,需对此 2 位进行置位操作。

编写单片机异步通信程序步骤如下：

(1) 设置串口工作方式。此时需对 SCON 中的 SM0、SM1 进行设置。PC 机与单片机的通信中一般选择串口工作在方式 1 下。

(2) 选择波特率发生器。选择定时器 1 或定时器 2 作为其波特率发生器。

(3) 设置定时器工作方式。当选择定时器 1 作为波特率发生器时,需设置其方式寄存器 TMOD 为计数方式并选择相应的工作方式（一般选择方式 2 以避免重装定时器初值）；当选择定时器 2 作为波特率发生器时,需将 T2CON 设置为波特率发生器工作方式。

(4) 设置波特率参数。影响波特率的参数有二,一是特殊寄存器 PCON 的 SMOD 位,另一个是相应定时器初值。

(5) 允许串行中断。因在程序中我们一般采有中断接收方式,故应设 EA=1,ES=1。

(6) 允许接收数据。设置 SCON 中的 REN 为 1,表示允许串行口接收数据。

(7)允许定时/计数器工作。此时开启定时/计数器,使其产生波特率。

(8)编写串行中断服务程序。当有数据到达串口时,系统将自动执行所编写的中断服务程序。

(9)收/发相应数据。注意的是发送操作完成需将 T1 清零,接收工作完成后需将 R1 清零。

2. 单片机与 PC 机异步串行通信

在微机中常用的有 9 针通信端口,在 PC 机中经常见到。基本的数据传送信号有:TXD(发送数据)、RXD(接收数据)、GND(地)。在 PC 机与单片机的通信中,一般只采用 3 根基本的数据传输线直接相连,另外,RS－232C 标准使用负逻辑。逻辑 1 的电平在－5V 到－15V 范围内,逻辑 0 的电平在＋5V 到＋15V 之间。这就是说 PC 机串行口的 TXD、RXD 信号线是不能与具有 TTL 电平的单片机信号线相连的。它们之间必须进行电平转换。因此常采用 RS－232C2;标准进行点对点的通信连接,信号采用 RS－232C 电平传输,电平转换芯片采用 MAX232。由通信的协议不同,它又可以分为单工通信和全双工通信。但无论是何种情况下,两个单片机的波特率必须一致,且工作于同一种串行方式下。下面就介绍单片机通过 9 针的 RS－232 实现串行异步通信。

单片机与 PC 机异步串行通信的基本编程方法和单片机与单片机异步串行通信的基本编程方法基本相同,首先要求制定一个双方都遵循的通信协议和波特率,另外在调试的时候可以使用一些串口调试工具来分别调试单片机程序和 PC 端的程序。

三、程序设计

1. 主函数设计

主程序主要完成数据初始化、硬件初始化、函数调用等功能。

(1)初始化

首先初始化串口成功接收数据标志位 recokbit 和串口接收数据单元 recdata 为 0。

然后调用 LCD 初始化函数,并设置 LCD 的 DDRAM 地址为 0x00,在 LCD 上显示数据 "RECDATA:"。

(2)串口设置

选择定时器 T1 作为波特率发生器,设定定时器 T1 的工作方式为方式 2,所以寄存器 TMOD 的初值应该为 0x20。

设置寄存器 SCON 的 SM0、SM1 位,定义串口工作方式为方式 1,并允许串口接收数据,定义 REN=1,所以寄存器 SCON 的初值应该为 0x50。

设置波特率参数为 9600bps,单片机的晶振 $fOSC=11.0592MHz$,$SMOD=0$,则定时器 T1 初值 X 为:$X=256-f_{osc}\times(SMOD+1)/(384\times 波特率)=253=0xfd$,所以 $TH1=TL1=0xfd$。

启动定时/计数器 T1 工作,所以 TR1=1。

开放总中断及串行中断,所以寄存器 IE 的初值应该为 0x90。

(3)串口收发数据

判断串口成功接收数据标志位 recokbit 是否为 0。

若 recokbit 为 0,表明串口未接收到数据,则继续等待串口接收数据。

若 recokbit 为 1,表明串口成功接收或发送数据,进入串口中断服务函数,单片机接收数据,并将串口成功接收数据标志位 recokbit 清零,调用 LCD 显示接收数据函数,在 LCD 上显示单片机从串口接收到的数据。

主函数设计流程图如图 21-2 所示。

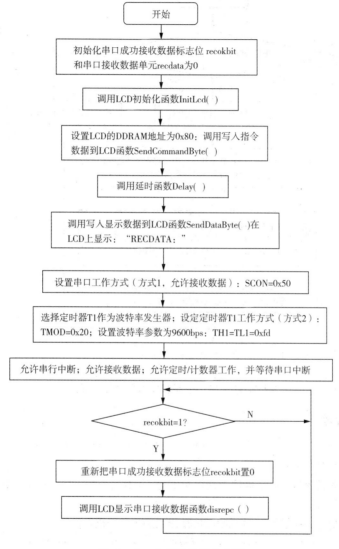

图 21-2 主函数设计流程图

2. 串口中断服务函数设计

首先判断串口接收中断标志位 RI 是否为 1。

(1)若 RI 不为 1,则串口发送中断标志位 TI=1,表示把数据从单片机串口发给 PC 机成功,并把串口发送中断标志位 TI 清零,串口中断函数返回。

(2)若 RI 为 1,表示单片机串口接收数据成功,则把串口接收中断标志位 RI 清零,把串口接收缓冲器 SBUF 中的数据写入串口接收数据单元 recdata。再把该数据送到串口发送缓冲器 SBUF 中,并回传给 PC 机,置串口成功接收数据标志位 recokbit 为 1,表明串口成功接收数据。

最后串口中断函数返回。

串口中断服务函数设计流程图如图 21-3 所示。

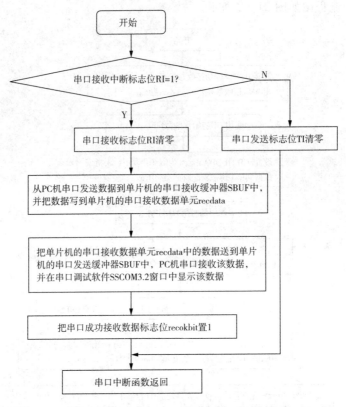

图 21-3 串口中断服务函数设计流程图

四、汇编源程序

```
;*******************************************************************************
;项目名称:RS-232 串行通信
;功能:单片机系统接收 PC 机串口发送过来的 ASCII 码,并在 LCD 上显示;同时单片机把接收到的 ASCII
;码回传给 PC 机串口并显示,以验证串口接收、发送数据的正确性。
;*******************************************************************************

sbit E = P2^2;          //LCD 使能信号
sbit RW = P2^1;         //读/写选择信号 R/W:0 为写入数据;1 为读出数据
sbit RS = P2^0;         //数据/命令选择信号 R/S:0 为指令;1 为数据

;*******************************************************************************
;功能:LCD1602 信号接口定义
;*******************************************************************************
E           BIT     P2.2        ;LCD 使能信号
RW          BIT     P2.1        ;LCD 读/写选择信号 R/W:0 为写入数据;1 为读出数据
RS          BIT     P2.0        ;LCD 数据/命令选择信号 R/S:0 为指令;1 为数据
```

任务21 RS-232 串行通信

```
        LCDPORT    EQU    P0
        CMD_BYTE   EQU    2EH
        DAT_BYTE   EQU    2FH

        RECOKBIT   BIT    00H        ;串口成功接收数据标志位
        RECDATA    EQU    30H        ;串口接收数据单元

        ORG        0000H
        AJMP       MAIN
```

;**
;功能:串口接收发送数据中断子程序
;**

```
        ORG        0023H              ;串口中断入口地址 0023H
RS232:  JB         TI,SEND232         ;若 TI=1,发送数据
        CLR        RI                 ;若 TI=0,则 RI=1,接收数据,并把 RI 清零
        MOV        A,SBUF             ;接收数据:PC—》SBUF—》A—》RECDATA
        MOV        RECDATA,A
        MOV        SBUF,A             ;发送数据:A—》SBUF—》PC
        SETB       RECOKBIT           ;把串口成功接收数据标志位 RECOKBIT 置 1
        RETI                          ;串口接收数据中断子程序返回
SEND232:CLR        TI                 ;TI=1,发送数据,并把 TI 清零
        RETI                          ;串口发送数据中断子程序返回
```

;**
;功能:主程序
;**

```
        ORG        0050H
MAIN:   MOV        20H,#00H           ;把串口成功接收数据标志位 RECOKBIT 置 0
        LCALL      INITLCD            ;调用 LCD 初始化子程序
        MOV        CMD_BYTE,#80H      ;设置 DDRAM 的地址
        LCALL      WRITE_CMD          ;调用写入指令数据到 LCD1602 子程序
        LCALL      DELAY0             ;调用延时子程序
        MOV        DAT_BYTE,#"R"
        LCALL      WRITE_DAT          ;调用写入显示数据到 LCD1602 子程序
        MOV        DAT_BYTE,#"E"
        LCALL      WRITE_DAT          ;调用写入显示数据到 LCD1602 子程序
        MOV        DAT_BYTE,#"C"
        LCALL      WRITE_DAT          ;调用写入显示数据到 LCD1602 子程序
        MOV        DAT_BYTE,#"D"
        LCALL      WRITE_DAT          ;调用写入显示数据到 LCD1602 子程序
        MOV        DAT_BYTE,#"A"
        LCALL      WRITE_DAT          ;调用写入显示数据到 LCD1602 子程序
```

· 177 ·

```
                MOV     DAT_BYTE,#"T"
                LCALL   WRITE_DAT           ;调用写入显示数据到LCD1602子程序
                MOV     DAT_BYTE,#"A"
                LCALL   WRITE_DAT           ;调用写入显示数据到LCD1602子程序
                MOV     DAT_BYTE,#":"
                LCALL   WRITE_DAT           ;调用写入显示数据到LCD1602子程序
                MOV     RECDATA,#00H
                MOV     TMOD,#20H           ;定时器1作为波特率发生器,工作模式为2
                MOV     TH1,#0FDH           ;波特率9600bps
                MOV     TL1,#0FDH
                MOV     SCON,#50H           ;串口方式1,允许接收数据
                SETB    TR1
                SETB    REN
                MOV     IE,#90H             ;允许串口中断
WAIT:           JNB     RECOKBIT,WAIT       ;等待串口中断
                CLR     RECOKBIT            ;把串口成功接收数据标志位RECOKBIT置0
                ACALL   DISPREC             ;调用LCD显示接收数据子程序
                SJMP    WAIT
```

;**
;功能:LCD1602要用到的一些子程序
;**
;**
;功能:写入指令数据到LCD1602子程序(入口参数 CMD_BYTE)
;**

```
WRITE_CMD:      CLR     RS
                CLR     RW
                MOV     A,CMD_BYTE
                MOV     LCDPORT,A
                SETB    E
                NOP
                NOP
                CLR     E
                LCALL   DELAY0
                RET                         ;写入指令数据到LCD1602子程序返回
```

;**
;功能:写入显示数据到LCD1602子程序(入口参数 DAT_BYTE)
;**

```
WRITE_DAT:      SETB    RS
                CLR     RW
                MOV     A,DAT_BYTE
                MOV     LCDPORT,A
```

```
              SETB     E
              NOP
              NOP
              CLR      E
              LCALL    DELAY0
              RET                           ;写入显示数据到 LCD1602 子程序返回

;****************************************************************************
;功能:LCD 初始化子程序
;****************************************************************************
INITLCD: MOV  CMD_BYTE,#30H
              LCALL    WRITE_CMD           ;调用写入指令数据到 LCD1602 子程序
              MOV      CMD_BYTE,#30H
              LCALL    WRITE_CMD           ;调用写入指令数据到 LCD1602 子程序
              MOV      CMD_BYTE,#30H
              LCALL    WRITE_CMD           ;调用写入指令数据到 LCD1602 子程序
              MOV      CMD_BYTE,#38H       ;设定工作方式
              LCALL    WRITE_CMD           ;调用写入指令数据到 LCD1602 子程序
              MOV      CMD_BYTE,#0CH       ;显示状态设置
              LCALL    WRITE_CMD           ;调用写入指令数据到 LCD1602 子程序
              MOV      CMD_BYTE,#01H       ;清屏
              LCALL    WRITE_CMD           ;调用写入指令数据到 LCD1602 子程序
              MOV      CMD_BYTE,#06H       ;输入方式设置
              LCALL    WRITE_CMD           ;调用写入指令数据到 LCD1602 子程序
              RET                          ;LCD 初始化子程序返回

;****************************************************************************
;功能:延时子程序
;****************************************************************************
DELAY0:  MOV  R5,#0A0H
DELAY1:  NOP
              DJNZ     R5,DELAY1
              RET                          ;子程序返回

;****************************************************************************
;功能:LCD 显示接收数据子程序 DISPREC
;****************************************************************************
DISPREC: MOV  CMD_BYTE,#89H                ;设置 DDRAM 的地址
              LCALL    WRITE_CMD           ;调用写入指令数据到 LCD1602 子程序
              MOV      DAT_BYTE,RECDATA
              LCALL    WRITE_DAT           ;调用写入显示数据到 LCD1602 子程序
              RET                          ;LCD 显示接收数据子程序返回
```

END

五、C语言源程序

```
//*********************************************************************
//项目名称:RS-232 串行通信
//功能:单片机系统接收 PC 机串口发送过来的 ASCII 码,并在 LCD 上显示;同时单片机把接收到的 ASCII
//码回传给 PC 机串口并显示,以验证串口接收、发送数据的正确性。
//*********************************************************************
#include <reg51.h>              //包含"51 寄存器定义"头文件

//*********************************************************************
//LCD1602 信号接口定义
//*********************************************************************
sbit E = P2^2;                  //LCD 使能信号
sbit RW = P2^1;                 //读/写选择信号 R/W:0 为写入数据;1 为读出数据
sbit RS = P2^0;                 //数据/命令选择信号 R/S:0 为指令;1 为数据
//*********************************************************************

//功能:变量定义
//*********************************************************************
unsigned char recdata;          //串口 RS232 接收的数据
unsigned char bdata myflag;     //定义一个可位寻址的片内数据存储单元
sbit recokbit = myflag^0;       //串口成功接收数据标志位

//*********************************************************************
//功能:函数声明
//*********************************************************************
void rs232(void);               //串口中断服务函数 rs232()
void Delay(unsigned int t);     //延时函数
void InitLcd(void);             //LCD1602 初始化函数 InitLcd()
void SendCommandByte(unsigned char ch);//写入指令数据到 LCD1602 函数
void SendDataByte(unsigned char ch); //写入显示数据到 LCD1602 函数
void disprec(void);             //LCD 显示串口接收数据函数 disrepc()

//*********************************************************************
//功能:主函数
//*********************************************************************
void main(void)
{
    myflag = 0x00;              //初始化串口成功接收数据标志位 recokbit 为 0
    recdata = 0x00;             //初始化串口接收数据单元 recdata 为 0
    InitLcd();                  //调用 LCD1602 初始化函数
```

```c
        SendCommandByte(0x80);     //调用写入指令数据到 LCD1602 函数
        Delay(2);                  //调用延时函数 Delay()
        SendDataByte('R');         //调用写入显示数据到 LCD1602 函数
        SendDataByte('E');         //调用写入显示数据到 LCD1602 函数
        SendDataByte('C');         //调用写入显示数据到 LCD1602 函数
        SendDataByte('D');         //调用写入显示数据到 LCD1602 函数
        SendDataByte('A');         //调用写入显示数据到 LCD1602 函数
        SendDataByte('T');         //调用写入显示数据到 LCD1602 函数
        SendDataByte('A');         //调用写入显示数据到 LCD1602 函数
        SendDataByte(':');         //调用写入显示数据到 LCD1602 函数
        TMOD = 0x20;               //定时器 1 作为波特率发生器,工作模式为 2
        TH1 = 0xfd;                //串行通信波特率为 9600 时,T1 初值:TH1 = 0xfd
        TL1 = 0xfd;                //串行通信波特率为 9600 时,T1 初值:TL1 = 0xfd
        SCON = 0x50;               //串口方式 1,允许接收数据
        TR1 = 1;                   //启动定时器 T1
        REN = 1;                   //允许串口接收数据
        IE = 0x90;                 //开放总中断和串口中断
        while(1)                   //等待串口接收或发送数据中断
        {
            if(recokbit = = 1)     //判断串口是否成功接收数据
            {
              recokbit = 0;        //重新把串口成功接收数据标志位 recokbit 置 0
              disprec();           //调用 LCD 显示串口接收数据函数 disrepc()
            }
        }
}


// ****************************************************************************
//功能:串口中断函数 rs232()
// ****************************************************************************
void rs232(void) interrupt 4
{
    //判断串口接收中断标志位 RI 是否为 1
    if(RI = = 1)        //RI 为 1,因串口接收数据而引起的串口中断
    {
       RI = 0;          //把串口接收中断标志位 RI 置 0
       recdata = SBUF;  //把接收缓冲器 SBUF 数据送给串口接收数据单元 recdata
       SBUF = recdata;  //把串口接收数据单元 recdata 数据送给发送缓冲器 SBUF
       recokbit = 1;    //把串口成功接收数据标志位 RECOKBIT 置 1
    }
    else                //RI 不为 1,因串口发送数据而引起的串口中断
    { TI = 0;}          //把串口发送中断标志位 TI 置 0
}
```

```c
//****************************************************************
//功能:延时函数 Delay()
//****************************************************************
Void Delay(unsigned int t)        //
{
    for(;t!=0;t--);
}


//****************************************************************
//功能:LCD1602 初始化函数 InitLcd( )
//****************************************************************
void InitLcd( )
{
    SendCommandByte(0x30);    //显示模式设置(不测试忙信号)共三次
    SendCommandByte(0x30);    //显示模式设置(不测试忙信号)共三次
    SendCommandByte(0x30);    //显示模式设置(不测试忙信号)共三次
    SendCommandByte(0x38);    //设置工作方式:8 位数据口,2 行显示,5*7 点阵
    SendCommandByte(0x08);    //设置显示状态:显示关
    SendCommandByte(0x01);    //清屏
    SendCommandByte(0x06);    //设置输入方式:数据读写后 AC 自增 1,画面不动
    SendCommandByte(0x0c);    //设置显示状态:显示开
}


//****************************************************************
//功能:写入指令数据到 LCD1602 函数 SendCommandByte()
//形参:ch
//****************************************************************
void SendCommandByte(unsigned char ch)
{
    RS = 0;                  //数据/命令选择信号 R/S:0 为指令信号;1 为数据信号
    RW = 0;                  //读/写选择信号 R/W:0 为写入数据;1 为读出数据
    P0 = ch;                 //把指令数据 ch 送至 P0 口(LCD 数据线 DB7~DB0)
    E = 1;                   //LCD 使能信号 E 置高电平
    Delay(1);                //延时 0.40us
    E = 0;                   //LCD 使能信号 E 置低电平
    Delay(100);              //delay40us
}


//****************************************************************
//功能:写入显示数据到 LCD1602 函数 SendDataByte()
//形参:ch
//****************************************************************
```

```c
{
    TH1 = 0xe0;
    TL1 = 0x00;
    timer1cnt++;
    if(timer1cnt>=(OSFREQ/1500000l))
        { timer1cntflg = 1;    TR1 = 0;}
}

void delay(uchar time)
{
    uchar i;
    uint j;
    for(i=0;i<time;i++)
      for(j=0;j<0x900;j++);
    /***
    uchar i;
    for(i=0;i<time;i++)
       {   timer1cnt = 0;   timer1cntflg = 0;
         TR1 = 1;
         while(! timer1cntflg);
       }
     ***/
}

void Sound(uint freq)
{
    uint timreg;
    timreg = 65536 - (OSFREQ/(25 * freq));
    FreqSandH = timreg/256;
    FreqSandL = timreg&0x00ff;
    TR0 = 1;   ET0 = 1;
}

void SoundOff(void)
{
    TR0 = 0;
    ET0 = 0;
    BEEP_PWR = 0;
}

uint GetFreq(uchar ch,uchar flg)
{
    uchar * pn,i=0;
```

```
void SendDataByte(unsigned char ch)
{
    RS = 1;              //数据/命令选择信号 R/S:0 为指令信号;1 为数据信号
    RW = 0;              //读/写选择信号 R/W:0 为写入数据;1 为读出数据
    P0 = ch;             //把显示数据 ch 送至 P0 口(LCD 数据线 DB7～DB0)
    E = 1;               //LCD 使能信号 E 置高电平
    Delay(1);            //延时 0.40us
    E = 0;               //LCD 使能信号 E 置低电平
    Delay(100);          //delay 40us
}

//*********************************************************************
//功能:LCD1602 显示串口接收数据函数 disprec()
//*********************************************************************
void disprec(void)
{
    SendCommandByte(0x89);      //调用写入指令数据到 LCD1602 函数
    SendDataByte(recdata);      //调用写入显示数据到 LCD1602 函数
}
```

六、系统调试

在 PC 机上打开串口调试软件——串口调试助手 SSCOM3.2,设置串口号、波特率、校验位等参数(注意:这些参数应与单片机中串口参数设置一致),在字符输入框内输入要发送的 ASCII 码,点击发送。此时在 LCD1602 上显示出当前收到的 ASCII 码,同时在串口调试助手 SSCOM3.2 的接收窗口可以看到当前的回传 ASCII 码。PC 机与单片机串行通信的调试过程如图 21-4 所示。

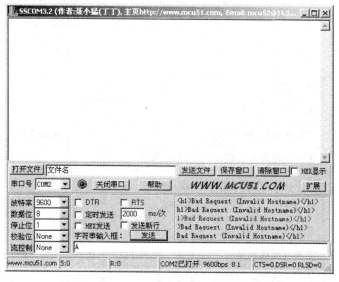

图 21-4　PC 机与单片机串行通信系统的调试示意图

实训考核

本课程改革传统的闭卷或开卷考核,而采用过程考核为主的多元化考核方式,考核分为理论考核、职业道德考核和技能考核三部分,各部分所占比例见表 21-2、表 21-3、表 21-4。

表 21-2 理论考核和职业素质考核形式及所占比例

序 号	名 称		比 例	得 分
一	理论考核	过程作业文件 个人自评	10%	
		过程作业文件 组内互评	20%	
		过程作业文件 小组互评	30%	
		过程作业文件 老师评定	20%	
		课堂提问、解答	10%	
		项目汇报	10%	
		小计	100%	
二	职业素质	职业道德,工作作风	40%	
		小组沟通协作能力	40%	
		创新能力	20%	
		小计	100%	

表 21-3 技能考核内容及比例

姓名		班级		小组		总得分	
序号	考核项目	考核内容及要求		配分	评分标准	考核环节	得分
1	①安全文明生产 ②安全操作规范	着装规范		20%	现场考评	实施	
		安全用电					
		布线规范、合理					
		工具摆放整齐					
		工具及仪器仪表使用规范、摆放整齐					
		任务完成后,进行场地整理,保持场地清洁、有序					
2	实训态度	不迟到、早退、旷课		10%	现场考评	六步	
		实训过程认真负责					
		组内主动沟通、协作,小组间互助					

(续表)

序号	考核项目	考核内容及要求	配分	评分标准	考核环节	得分
3	系统方案制定	工作流程正确合理	10%	现场考评	计划决策	
		方案合理				
		选用指令是否合理				
		电路图正确				
	编程能力	独立完成程序	10%	现场考评	决策	
		程序简单、可靠				
4	操作能力	正确输入程序并进行程序调试	20%	现场考评	实施	
		根据电路图正确接线				
		根据系统功能进行正确操作演示				
5	工艺	接线美观	10%	现场考评	实施	
		线路工作可靠				
6	实践效果	系统工作可靠	10%	现场考评	检查	
		满足工作要求				
		创新				
		按规定的时间完成项目				
7	汇报总结	工作总结,PPT汇报	5%	现场考评	评估	
		填写自我检查表及反馈表				
8	技术文件制作整理	技术文件制作整理能力	5%	现场考评	评估	
	合计		100%			

表 21-4 **各部分考核占课程考核的比例**

考核项目	理论考核	技能考核	职业素质考核	合 计
比 例	30%	50%	20%	100%
分 值	30	50	20	100
实际得分				

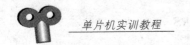

任务22 A/D 转换

实训任务

通过可调电位器 VR2 调节电压来模拟模拟电压的输入,范围在 0~5V,然后进行 A/D 转换,转换后的数字量即输出电压值通过 LCD 显示。

实训设备

1. 设备
PC 机(安装 wave 编程软件、Keil C51 软件)、单片机实验板。
2. 工具及材料
工作对象:电工电子工具、电子元器件和辅助材料、仿真器、编程器。
工作工具:单片机控制电路原理图、实训指导书、项目任务单、工作记录单、项目检查单、各种电工仪表、常用电工工具和拆装工具、量具、相关电子手册。

硬件设计

主控模块采用 ATMEL 公司生产的 AT89S52 单片机,LCD 显示模块选用 1602 字符型 LCD 模块。A/D 模块由 A/D 芯片和可调电位器电路组成,通过可调电位器电路实现模拟电压输入,范围在 0~5V 之间,A/D 芯片选用 TLC549 芯片。A/D 模块与单片机的接口电路如图 22-1 所示。

软件设计

一、A/D 转换器 TLC549

1. 芯片特点

使用串行 A/D 转换器 TLC549,该芯片的特点有:①用 CMOS 技术;②8 位转换结果;③与微处理器或外围设备接口;④差分基准电压输入;⑤转换时间:最大 17us;⑥每秒访问和转换次数:达到 40000;⑦片上软件控制采样和保持功能;⑧全部非校准误差:±0.5LSB;⑨宽电压供电:3~6V 封装及引脚;⑩低功耗:最大 15mW;⑪5V 供电时输入范围:0~5V;⑫输入输出完全兼容 TTL 和 CMOS 电路;⑬全部非校准误差:±1LSB;⑭工作温度范围:0℃~70℃(TLC549);-40℃~85℃(TLC549I)。

2. 工作时序

其工作时序如图 22-3 所示。当 CS 为高时,数据输出 DATA_OUT 端处于高阻状态,

任务 22　A/D 转换

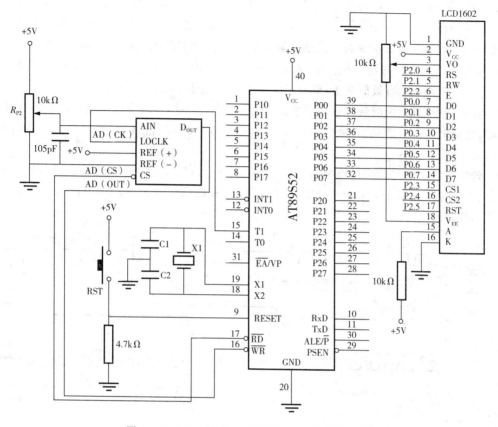

图 22-1　A/D 模块与单片机的接口电路原理图

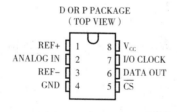

图 22-2　TLC549 引脚图

此时 DA_CLOCK 不起作用。这种 CS 控制作用允许在同时使用多片 TLC549 时,共用 DA_CLOCK,以减少多路(片)A/D 并用时的 I/O 控制端口。

一组通常的控制时序为:

(1)将 CS 置低。内部电路在测得 CS 下降沿后,再等待两个内部时钟上升沿和一个下降沿后,然后确认这一变化,最后自动将前一次转换结果的最高位(D7)位输出到 DATA OUT 端上。

(2)前四个 DA_CLOCK 周期的下降沿依次移出第 2、3、4 和第 5 个位(D6、D5、D4、D3),片上采样保持电路在第 4 个 DA_CLOCK 下降沿开始采样模拟输入。

(3)接下来的 3 个 DA_CLOCK 周期的下降沿移出第 6、7、8(D2、D1、D0)个转换位。

(4)最后,片上采样保持电路在第 8 个 I/O CLOCK 周期的下降沿将移出第 6、7、8(D2、

D1、D0)个转换位。保持功能将持续 4 个内部时钟周期，然后开始进行 32 个内部时钟周期的 A/D 转换。第 8 个 DA_CLOCK 后，CS 必须为高，或 DA_CLOCK 保持低电平，这种状态需要维持 36 个内部系统时钟周期以等待保持和转换工作的完成。如果 CS 为低时 DA_CLOCK 上出现一个有效干扰脉冲，则微处理器/控制器将与器件的 I/O 时序失去同步；若 CS 为高时出现一次有效低电平，则将使引脚重新初始化，从而脱离原转换过程。

在 36 个内部系统时钟周期结束之前，实施步骤(1)~(4)，可重新启动一次新的 A/D 转换，与此同时，正在进行的转换终止，此时的输出是前一次的转换结果而不是正在进行的转换结果。

若要在特定的时刻采样模拟信号，应使第 8 个 I/O CLOCK 时钟的下降沿与该时刻对应，因为芯片虽在第 4 个 I/O CLOCK 时钟下降沿开始采样，却在第 8 个 DA_CLOCK 下降沿开始保存。（注：在上升沿时读数据。）

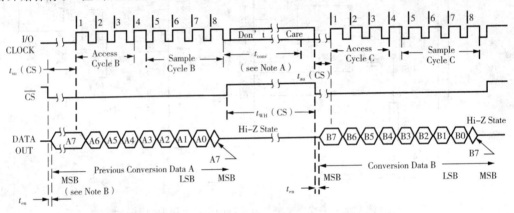

图 22-3　TLC549 工作时序

二、算法设计

通过可调电位器 RP2 改变电压输出值在 0~5V 范围内连续变化，从而可以模拟模拟量的输入。

根据 A/D 转换芯片 TLC549 的工作时序，20ms 进行一次 A/D 采样转换，可以利用定时器 T0 定时，基本定时时间为 5ms，控制软计数器的累计次数为 4 次，20ms(4×5ms)定时到时，产生定时器 T0 中断，在定时器 T0 中断服务函数中调用 AD 转换函数进行 A/D 采样转换，然后调用计算 A/D 转换值函数把 A/D 转换值转换为相应的 ASCII 码，最后通过 LCD 显示 A/D 转换值函数把输出电压(0~5V)的转换数字量显示在液晶 LCD1602 上，数字量显示值范围为 0~255。

三、程序设计

1. 主函数设计

主函数主要完成硬件初始化、数据初始化、函数调用等功能。

(1)初始化

首先初始化定时软计数器 keytime 值为 4。调用 LCD 初始化函数，调用写入显示数据到 LCD1602 函数设置 LCD 的 DDRAM 地址为 00H，调用延时函数，调用写入显示数据到 LCD1602 函数在 LCD 上显示字符数据"ADC:"。

(2)定时初值计算

定时器 T0 的定时时间为 5ms,系统所用的石英晶体振荡频率为 11.0592MHz,因此,1个机器周期=1/石英频率×12,即为 12/11.0592us,定时器的工作方式设置为方式 1,计算初值如下:

$$X = 2^{16} - t \times f_{osc}/12 = 65536 - 5 \times 10^{-3} \times 11.0592 \times 10^6/12 = 60928 = 0xEE00$$

所以 TH0=0xee,TL0=0x00。

(3)定时器设置

设定定时器 T0 工作方式:定时器、方式 1,即 TMOD=0x01。

启动定时器 T0,即 TR0=1。

开放定时器 T0 中断以及总中断,即设定 IE=0x82。

(4)等待中断

定时器 T0 启动计时后,CPU 等待定时中断的到来。当定时器 T0 定时 5ms 后,进入定时器 T0 中断服务函数。

主函数设计流程图如图 22-4 所示。

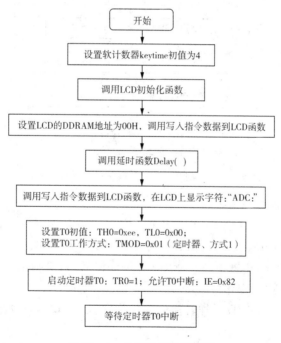

图 22-4 主函数设计流程图

2. 定时器 T0 中断服务函数模块设计

当定时器 T0 定时 5ms 后,进入定时器 T0 中断服务函数。

首先重装定时器 T0 初值,即 TH0=0xee,TL0=0x00。每定时 5ms 一次,软计数器 KEYTIME 值减 1。然后判断软计数器 keytime 值是否为 0。

(1)若 keytime 值不为 0,表明 20ms(20ms 采样一次)计时未到,这时 T0 中断函数返回主函数,继续计时。

(2)若 keytime 值为 0,表明 20ms(20ms 采样一次)计时已到,重置软计数器 keytime 初值为4,为下次定时做准备,接着调用 A/D 转换函数进行 A/D 采样转换,得到 A/D 采样转换值 adbl,然后调用计算 A/D 转换值 ASCII 码函数计算 A/D 采样转换值 adbl 相应的 ASCII 码,再调用 LCD 显示 A/D 转换值函数把模拟量输入电压(0~5V)的转换数字量显示在液晶 LCD1602 上,数字量显示值范围为 0~255。

最后 T0 中断函数返回主函数进行下一次 A/D 采样转换。

定时器 T0 中断服务函数设计流程图如图 22-5 所示。

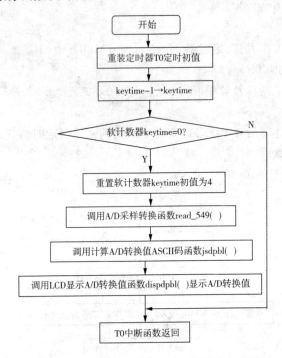

图 22-5 定时器 T0 中断服务函数设计流程图

3. A/D 采样转换函数模块设计

根据 A/D 转换芯片 TLC549 的工作时序,当片选信号 CS 为高电平时,数据输出 DATA OUT 端处于高阻状态,此时时钟信号 I/O CLOCK 不起作用,不能进行 A/D 转换。将片选信号 CS 置低电平,内部电路在测得 CS 下降沿后,再等待两个内部时钟上升沿和一个下降沿后,然后确认这一变化。

首先定义变量 i(表示 A/D 采样转换位数),并将芯片 TLC549 片选信号 AD_CS 置低电平,选中该芯片。

初始化 A/D 采样转换值 adbl 为 0,初始化变量 i(A/D 采样转换位数)为 0。

判断 A/D 转换位数 i 是否小于 8。

(1)如果 i 小于 8,则将芯片 TLC549 时钟信号 AD_CK 置高电平,并把 A/D 采样转换值 adbl 左移一位。然后判断 A/D 采样转换串行数据输出信号 AD_OUT 是否为 1:若 AD_OUT 为 1,则将 A/D 采样转换值 adbladbl 自加 1;若 AD_OUT 为 0,则 A/D 采样转换值 adbladbl 不变。

再将芯片 TLC549 时钟信号 AD_CK 置低电平。最后将 A/D 转换位数 i 自加 1,并再次转向上面的判断 A/D 转换位数 i 是否小于 8。

(2)如果 i 不小于 8,则把芯片 TLC549 片选信号 AD_CS 置高电平,结束 A/D 采样转换,并退出 A/D 采样转换函数。

A/D 采样转换函数设计流程图如图 22-6 所示。

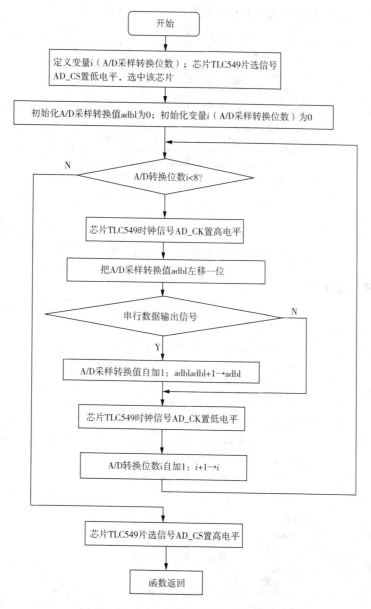

图 22-6 A/D 采样转换函数设计流程图

4. 计算 A/D 转换值 ASCII 码函数模块设计

要把 AD 转换值 adbl 显示在 LCD 上,需要把它转换为相应的 ASCII 码。

首先计算 A/D 转换值百位数 ASCII 码:将 A/D 转换值 adbl 除以 100 得到的商与 0x30(因为字符数字 0~9 与其相应的 ASCII 码相差 30H)相与。

然后计算 A/D 转换值十位数 ASCII 码:将 A/D 转换值 adbl 除以 100 得到的余数再除以 10,得到的商与 0x30(因为字符数字 0~9 与其相应的 ASCII 码相差 30H)相与。

再计算 A/D 转换值个位数 ASCII 码:将 A/D 转换值 adbl 除以 10 得到的余数与 0x30 (因为字符数字 0~9 与其相应的 ASCII 码相差 30H)相与。

最后函数返回。

计算 A/D 转换值 ASCII 码函数设计流程图如图 22-7 所示。

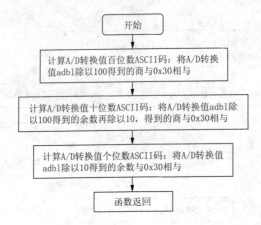

图 22-7　计算 A/D 转换值 ASCII 码函数设计流程图

四、汇编语言源程序

```
;******************************************************************
;项目名称:A/D 转换
;功能:在 LCD1602 上显示当前模拟电压(0~5V)的 A/D 转换值 000~255
;******************************************************************


;******************************************************************
;功能:LCD1602 与单片机信号接口定义
;******************************************************************
E           BIT     P2.2        ;LCD 使能信号
RW          BIT     P2.1        ;读/写选择信号 R/W:0 为写入数据;1 为读出数据
RS          BIT     P2.0        ;数据/命令选择信号 R/S:0 为指令;1 为数据
LCDPORT     EQU     P0          ;LCD1602 数据线 DB7~DB0
CMD_BYTE    EQU     2EH         ;写指令数据入口参数
DAT_BYTE    EQU     2FH         ;写显示数据入口参数


;******************************************************************
;功能:A/D 芯片 TLC549 与单片机接口定义
;******************************************************************
AD_CS       BIT     P3.7        ;芯片 TLC549 的片选信号
AD_CK       BIT     P3.5        ;芯片 TLC549 的时钟信号
AD_OUT      BIT     P3.6        ;芯片 TLC549 的串行数据输出信号


;******************************************************************
;功能:
```

任务22 A/D 转换

```
;*************************************************************
ADBL        EQU     30H         ;当前 AD 的变量(00~255 之间)
LEDBAI      EQU     31H         ;显示的百位
LEDSHI      EQU     32H         ;显示的十位
LEDGE       EQU     33H         ;显示的个位

            ORG     0000H
            AJMP    MAIN
            ORG     000BH       ;定时器 T0 中断服务入口地址
            AJMP    TIME0_1

;*************************************************************
;功能:主程序
;*************************************************************
            ORG     0030H
MAIN:       MOV     20H,#00H
            LCALL   INITLCD     ;调用 LCD 初始化子程序
            MOV     ADBL,#00H   ;AD 的变量清零
            MOV     CMD_BYTE,#80H ;设置 DDRAM 的地址
            LCALL   WRITE_CMD   ;调用写入指令数据到 LCD1602 子程序
            LCALL   DELAY0      ;调用延时子程序
            MOV     DAT_BYTE,#"A"
            LCALL   WRITE_DAT   ;调用写入显示数据到 LCD1602 子程序
            MOV     DAT_BYTE,#"D"
            LCALL   WRITE_DAT   ;调用写入显示数据到 LCD1602 子程序
            MOV     DAT_BYTE,#"C"
            LCALL   WRITE_DAT   ;调用写入显示数据到 LCD1602 子程序
            MOV     DAT_BYTE,#":"
            LCALL   WRITE_DAT   ;调用写入显示数据到 LCD1602 子程序
            MOV     TH0,#0EEH   ;16 位的定时器,定时 5ms,晶振为 11.0592MHz
            MOV     TL0,#00H
            MOV     TMOD,#01H
            SETB    TR0
            MOV     IE,#82H     ;允许定时器 T0 中断
            SJMP    $           ;等待定时器 T0 中断

;*************************************************************
;功能:定时器 T0 中断服务子程序
;*************************************************************
            ORG     0100H
TIME0_1:    MOV     TH0,#0EEH
            MOV     TL0,#00H
            DJNZ    KEYTIME,TIME0_RE
```

```
            MOV     KEYTIME,#04H        ;每 20ms 进行一次按键扫描
            LCALL   ADC549              ;调用 A/D 转换子程序
            LCALL   JSDPBL              ;调用计算 A/D 值子程序
            LCALL   DISPDPBL            ;调用 LCD 显示 A/D 值子程序
TIME0_RE：  RETI                        ;中断子程序返回
```

;***
;功能:A/D 转换子程序
;***

```
ADC549：    CLR     AD_CS
            MOV     R7,#08H
            MOV     ADBL,#00H
READ_1：    SETB    AD_CK
            JB      AD_OUT,SETBC
            CLR     C
            AJMP    READ_2
SETBC：     SETB    C
READ_2：    MOV     A,ADBL
            RLC     A
            MOV     ADBL,A
            CLR     AD_CK
            DJNZ    R7,READ_1
            SETB    AD_CS
            RET
```

;***
;功能:LCD1602 要用到的一些子程序
;***
;功能:写入指令数据到 LCD 子程序(入口参数 CMD_BYTE)
;***

```
WRITE_CMD:  CLR     RS
            CLR     RW
            MOV     A,CMD_BYTE
            MOV     LCDPORT,A
            SETB    E
            NOP
            NOP
            CLR     E
            LCALL   DELAY0
            RET
```

;***

;功能:写入显示数据到 LCD 子程序(入口参数 DAT_BYTE)
;**
WRITE_DAT: SETB   RS
           CLR   RW
           MOV   A,DAT_BYTE
           MOV   LCDPORT,A
           SETB   E
           NOP
           NOP
           CLR   E
           LCALL   DELAY0
           RET

;**
;功能:LCD 显示初始化子程序
;**
INITLCD: MOV   CMD_BYTE,#30H
           LCALL   WRITE_CMD   ;调用写入指令数据到 LCD1602 函数
           MOV   CMD_BYTE,#30H
           LCALL   WRITE_CMD   ;调用写入指令数据到 LCD1602 函数
           MOV   CMD_BYTE,#30H
           LCALL   WRITE_CMD   ;调用写入指令数据到 LCD1602 函数
           MOV   CMD_BYTE,#38H   ;设定工作方式
           LCALL   WRITE_CMD   ;调用写入指令数据到 LCD1602 函数
           MOV   CMD_BYTE,#0CH   ;显示状态设置
           LCALL   WRITE_CMD   ;调用写入指令数据到 LCD1602 函数
           MOV   CMD_BYTE,#01H   ;清屏
           LCALL   WRITE_CMD   ;调用写入指令数据到 LCD1602 函数
           MOV   CMD_BYTE,#06H   ;输入方式设置
           LCALL   WRITE_CMD   ;调用写入指令数据到 LCD1602 函数
           RET   ;子程序返回

;**
;功能:延时子程序
;**
DELAY0:   MOV   R5,#0A0H
DELAY1:   NOP
           DJNZ   R5,DELAY1
           RET

;**
;功能:计算 A/D 转换值子程序
;**

```
JSDPBL:   MOV    A,ADBL          ;把 A/D 转换值 ADBL 送给累加器 A
          MOV    B,#64H          ;给寄存器 B 赋值 100
          DIV    AB              ;除以 100,得到百位数字,商数存 A 余数存 B
          ORL    A,#30H          ;加 30H,变换成相应的 ASCII
          MOV    LEDBAI,A        ;送到 LCD 显示

          MOV    A,B             ;把 B 中的余数送给累加器 A
          MOV    B,#0AH          ;给寄存器 B 赋值 10
          DIV    AB              ;除以 10,得到十位数字,商数存 A 余数存 B
          ORL    A,#30H          ;加 30H,变换成相应的 ASCII
          MOV    LEDSHI,A        ;送到 LCD 显示

          MOV    A,B             ;余数为个位
          ORL    A,#30H          ;加 30H,变换成相应的 ASCII
          MOV    LEDGE,A         ;送到 LCD 显示
          RET                    ;子程序返回

;************************************************************************
;功能:LCD 显示 A/D 转换值子程序
;************************************************************************
DISPDPBL: MOV    CMD_BYTE,#85H   ;设置 DDRAM 的地址
          LCALL  WRITE_CMD       ;调用写入指令数据到 LCD1602 函数
          MOV    DAT_BYTE,LEDBAI ;LCD 显示 AD 转换值的百位
          LCALL  WRITE_DAT       ;调用写入显示数据到 LCD1602 函数
          MOV    DAT_BYTE,LEDSHI ;LCD 显示 AD 转换值的十位
          LCALL  WRITE_DAT       ;调用写入显示数据到 LCD1602 函数
          MOV    DAT_BYTE,LEDGE  ;LCD 显示 AD 转换值的个位
          LCALL  WRITE_DAT       ;调用写入显示数据到 LCD1602 函数
          RET                    ;子程序返回

          END
```

五、C 语言源程序

```
//****************************************************************************//
//项目名称:A/D 转换
//功能:在 LCD1602 上显示当前模拟电压(0~5V)的 A/D 转换值 000~255
//****************************************************************************//
#include <reg51.h>              //包含"51 寄存器定义"头文件


//****************************************************************************
//LCD1602 信号接口定义
//****************************************************************************
sbit E = P2^2;                  //LCD 使能信号
```

任务22 A/D 转换

```c
sbit RW = P2^1;                //读/写选择信号 R/W:0 为写入数据;1 为读出数据
sbit RS = P2^0;                //数据/命令选择信号 R/S:0 为指令信号;1 为数据信号

// ***************************************************************
// AD 芯片 TLC549 信号接口定义
// ***************************************************************
sbit AD_CK = P3^5;             //A/D 芯片 TLC549 时钟信号
sbit AD_OUT = P3^6;            //A/D 芯片 TLC549 数据输出信号
sbit AD_CS = P3^7;             //A/D 芯片 TLC549 片选信号

// ***************************************************************
//定义变量:
//adbl—当前的 AD 变量
//ledbai—数字量输出值百位显示值
//ledshi—数字量输出值十位显示值
//ledge—数字量输出值个位显示值
//keytime—软计数器
// ***************************************************************
unsigned char adbl,ledbai,ledshi,ledge,keytime;

// ***************************************************************
//功能:函数声明
// ***************************************************************
void time0(void);                          //定时器 T0 中断服务函数 time0()
void read_549(void);                       //A/D 转换函数 read_549()
void InitLcd(void);                        //LCD 初始化函数
void Delay(unsigned int t);                //延时函数 Delay()
void SendCommandByte(unsigned char ch);    //写入指令数据到 LCD1602 函数
void SendDataByte(unsigned char ch);       //写入显示数据到 LCD1602 函数
voidjsdpbl(void);                          //计算 A/D 转换值 ASCII 码函数 jsdpbl()
void dispdpbl(void);                       //LCD 显示 A/D 转换值函数 dispdpbl()

// ***************************************************************
//功能:主函数
// ***************************************************************
void main(void)                            //主函数入口地址
{
    keytime = 4;                           //初始化软计数器值为 4,20ms 扫描一次按键
    InitLcd();                             //调用 LCD 初始化函数 InitLcd()
    SendCommandByte(0x80);                 //设置 LCD 的 DDRAM 地址为 00H
    Delay(2);                              //调用延时函数
    SendDataByte('A');                     //调用写入显示数据到 LCD1602 函数
    SendDataByte('D');                     //调用写入显示数据到 LCD1602 函数
```

```c
        SendDataByte('C');              //调用写入显示数据到LCD1602函数
        SendDataByte(':');              //调用写入显示数据到LCD1602函数
        TH0 = 0xee;                     //定时器T0定时5ms初值设置:TH0 = 0xee
        TL0 = 0x00;                     //定时器T0定时5ms初值设置:TL0 = 0x00
        TMOD = 0x01;                    //定时器T0工作方式设置:定时器、方式1
        TR0 = 1;                        //启动定时器T0
        IE = 0x82;                      //开放总中断和定时器T0中断
        while(1)                        //等待定时器T0中断的到来
        { }
}


//**************************************************************************
//功能:定时器T0中断服务函数
//**************************************************************************
void time0(void) interrupt 1      //定时器T0中断服务函数
{
        TH0 = 0xee;                     //重新设置定时器T0初值:TH0 = 0xee
        TL0 = 0x00;                     //重新设置定时器T0初值:TL0 = 0x00
        keytime - - ;                   //每定时5ms一次,软计数器keytime值减1
        if(keytime = = 0)               //判断20ms定时时间是否到达,如达到,开始A/D采样
        {
                keytime = 4;            //重新置软计数器值为4,20ms采样一次
                read_549();             //调用A/D采样转换函数read_549()
                jsdpbl();               //调用计算A/D转换值ASCII码函数jsdpbl()
                dispdpbl();             //调用LCD显示A/D转换值函数dispdpbl( )
        }
}


//**************************************************************************
//功能:A/D采样转换函数read_549()
//**************************************************************************
void read_549(void)               //A/D采样转换函数
{
        unsigned char i;                //定义变量i:A/D转换位数
        AD_CS = 0;                      //芯片TLC549片选信号AD_CS置低电平
        adbl = 0x00;                    //初始化A/D采样转换值adbl为零
        for(i = 0;i<8;i + + )            //8位循环采样
        {
                AD_CK = 1;              //芯片TLC549时钟信号AD_CK置高电平
                adbl = adbl<<1;         //将A/D采样转换值adbl左移位一位
                                        //判断TLC549的串行数据输出信号AD_OUT = 1?
                if(AD_OUT = = 1)
                        adbl + + ;      //如AD_OUT为1,则adbl自加1
```

任务 22 A/D 转换

```
        AD_CK = 0;              //将芯片 TLC549 时钟信号 AD_CK 置低电平
    }
    AD_CS = 1;                  //将芯片 TLC549 片选信号 AD_CS 置高电平
}

//******************************************************************************
//功能:延时函数 Delay()
//******************************************************************************
void Delay(unsigned int t)
{
    for(;t! = 0;t- -) ;
}


//******************************************************************************
//功能: LCD1602 初始化函数 InitLcd( )
//******************************************************************************
void InitLcd( )
{
    SendCommandByte(0x30);   //显示模式设置(不测试忙信号)共三次
    SendCommandByte(0x30);   //显示模式设置(不测试忙信号)共三次
    SendCommandByte(0x30);   //显示模式设置(不测试忙信号)共三次
    SendCommandByte(0x38);   //设置工作方式:8 位数据口,2 行显示,5 * 7 点阵
    SendCommandByte(0x08);   //设置显示状态:显示关
    SendCommandByte(0x01);   //清屏
    SendCommandByte(0x06);   //设置输入方式:数据读写后 AC 自增 1,画面不动
    SendCommandByte(0x0c);   //设置显示状态:显示开
}


//******************************************************************************
//功能:写入指令数据到 LCD1602 函数 SendCommandByte()
//形参:ch
//******************************************************************************
void SendCommandByte(unsigned char ch)
{
    RS = 0;                 //数据/命令选择信号 R/S:0 为指令信号;1 为数据信号
    RW = 0;                 //读/写选择信号 R/W:0 为写入数据;1 为读出数据
    P0 = ch;                //把指令数据 ch 送至 P0 口(LCD 数据线 DB7~DB0)
    E = 1;                  //LCD 使能信号 E 置高电平
    Delay(1);               //延时 0.40us
    E = 0;                  //LCD 使能信号 E 置低电平
    Delay(100);             //delay40us
}
```

```
// ****************************************************************
//功能:写入显示数据到LCD1602函数 SendDataByte()
//形参:ch
// ****************************************************************
void SendDataByte(unsigned char ch)
{
    RS = 1;                    //数据/命令选择信号 R/S:0 为指令信号;1 为数据信号
    RW = 0;                    //读/写选择信号 R/W:0 为写入数据;1 为读出数据
    P0 = ch;                   //把显示数据 ch 送至 P0 口(LCD 数据线 DB7~DB0)
    E = 1;                     //LCD 使能信号 E 置高电平
    Delay(1);                  //延时 0.40us
    E = 0;                     //LCD 使能信号 E 置低电平
    Delay(100);                //delay 40us
}


// ****************************************************************
//功能:计算 A/D 转换值 ASCII 码函数 jsdpbl()
// ****************************************************************
void jsdpbl(void)
{
    //计算 A/D 转换值百位,并与 0x30 相与,得到 A/D 转换值百位的 ASCII 码
    ledbai = (adbl/100)|0x30;
    //计算 A/D 转换值十位,并与 0x30 相与,得到 A/D 转换值十位的 ASCII 码
    ledshi = ((adbl%100)/10)|0x30;
    //计算 A/D 转换值个位,并与 0x30 相与,得到 A/D 转换值个位的 ASCII 码
    ledge = (adbl%10)|0x30;
}


// ****************************************************************
//功能:LCD 显示 A/D 转换值函数 dispdpbl()
// ****************************************************************
void dispdpbl(void)              //LCD 显示 A/D 转换值函数
{
    SendCommandByte(0x85);       //设置 LCD 的 DDRAM 地址为 05H
    SendDataByte(ledbai);        //LCD 显示 A/D 转换值百位
    SendDataByte(ledshi);        //LCD 显示 A/D 转换值十位
    SendDataByte(ledge);         //LCD 显示 A/D 转换值个位
}
```

实训考核

本课程改革传统的闭卷或开卷考核,而采用过程考核为主的多元化考核方式,考核分为理论考核、职业道德考核和技能考核三部分,各部分所占比例见表 22-1、表 22-2、表 22-3

所示。

表22-1 理论考核和职业素质考核形式及所占比例

序号	名称		比例	得分
一	理论考核	个人自评	10%	
		过程作业文件 组内互评	20%	
		小组互评	30%	
		老师评定	20%	
		课堂提问、解答	10%	
		项目汇报	10%	
		小计	100%	
二	职业素质	职业道德,工作作风	40%	
		小组沟通协作能力	40%	
		创新能力	20%	
		小计	100%	

表22-2 技能考核内容及比例

姓名		班级		小组		总得分	
序号	考核项目	考核内容及要求		配分	评分标准	考核环节	得分
1	①安全文明生产 ②安全操作规范	着装规范		20%	现场考评	实施	
		安全用电					
		布线规范、合理					
		工具摆放整齐					
		工具及仪器仪表使用规范、摆放整齐					
		任务完成后,进行场地整理,保持场地清洁、有序					
2	实训态度	不迟到、早退、旷课		10%	现场考评	六步	
		实训过程认真负责					
		组内主动沟通、协作,小组间互助					
3	系统方案制定	工作流程正确合理		10%	现场考评	计划决策	
		方案合理					
		选用指令是否合理					
		电路图正确					
	编程能力	独立完成程序		10%	现场考评	决策	
		程序简单、可靠					

(续表)

序号	考核项目	考核内容及要求	配分	评分标准	考核环节	得分
4	操作能力	正确输入程序并进行程序调试	20%	现场考评	实施	
		根据电路图正确接线				
		根据系统功能进行正确操作演示				
5	工艺	接线美观	10%	现场考评	实施	
		线路工作可靠				
6	实践效果	系统工作可靠	10%	现场考评	检查	
		满足工作要求				
		创新				
		按规定的时间完成项目				
7	汇报总结	工作总结，PPT汇报	5%	现场考评	评估	
		填写自我检查表及反馈表				
8	技术文件制作整理	技术文件制作整理能力	5%	现场考评	评估	
	合计		100%			

表22-3 各部分考核占课程考核的比例

考核项目	理论考核	技能考核	职业素质考核	合 计
比 例	30%	50%	20%	100%
分 值	30	50	20	100
实际得分				

任务23 D/A 转换

实训任务

由 PC 机串口发送到单片机串口的值作为数字量初始值,通过按键 K1、K2 实现数字量的加减。按键 K1 实现数字量初始值加 1,按键 K2 实现数字量初始值减 1,使数字量在 20~200 的范围内变化,从而模拟数字量信号的输入。经过 D/A 转换,DAC 转换的模拟电压值从 DA 输出座子输出,可用万用表测量,并且模拟电压值在 LCD1602 上显示。

实训设备

1. 设备
PC 机(安装 wave 编程软件、Keil C51 软件)、单片机实验板。
2. 工具及材料
工作对象:电工电子工具、电子元器件和辅助材料、仿真器、编程器。
工作工具:单片机控制电路原理图、实训指导书、项目任务单、工作记录单、项目检查单、各种电工仪表、常用电工工具和拆装工具、量具、相关电子手册。

硬件设计

主控模块采用 ATMEL 公司生产的 AT89S52 单片机。LCD 显示模块选用 1602 字符型 LCD 模块。键盘输入使用独立式按键 K1、K2。串口通信电路模块使用 MAXIM 公司的 MAX232 集成芯片。D/A 芯片选用 DAC0832。D/A 模块与单片机的接口电路如图 23-1 所示。

软件设计

一、D/A 转换器 DAC0832

DAC0832 是目前国内使用较普遍的 D/A 转换器。

1. DAC0832 的主要特性

DAC0832 是采用 CMOS/Si—Cr 工艺制成的双列直插式单片 8 位 D/A 转换器。它可直接与 Z80,8085,8080 等 CPU 相连,也可与 8031 相连,以电流形式输出;当转换为电压输出时,可外接运算放大器。其主要特性有:

① 输出电流线性度可在满量程下调节。
② 转换时间为 $1\mu s$。

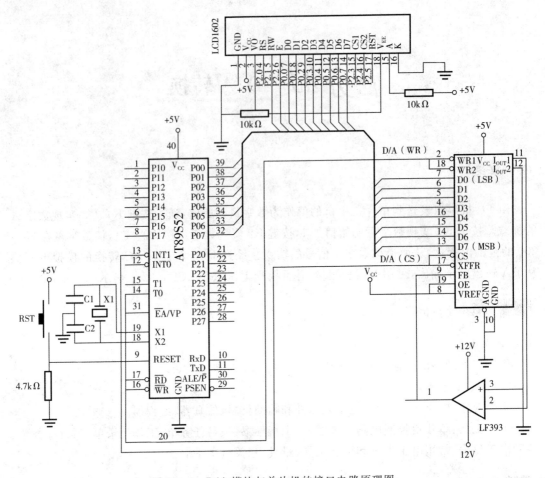

图 23-1 D/A 模块与单片机的接口电路原理图

③ 数据输入可采用双缓冲、单缓冲或直通方式。

④ 增益温度补偿为 0.02%FS/℃。

⑤ 每次输入数字为 8 位二进制数。

⑥ 功耗为 20mW。

⑦ 逻辑电平输入与 TTL 兼容。

⑧ 供电电源为单一电源,可在 5~15V 内。

2. DAC0832 的内部结构及外部引脚

DAC0832 型 D/A 转换器,其内部结构由一个数据寄存器、DAC 寄存器和 D/A 转换器三大部分组成。

DAC0832 内部采用 R-2R 梯形电阻网络。两个寄存器——输入数据寄存器和 DAC 寄存器用以实现两次缓冲,故在输出的同时,可集成一个数字,这就提高了转换速度。当多芯片同时工作时,可用同步信号实现各模拟量同时输出。DAC0832 的外部引脚如图 23-2 所示。

图 23-2 DAC0832 的引脚

$\overline{CS}$——片选信号,低电平有效。与 ILE 相配合,可对写信号$\overline{WR1}$是否有效起到控制作用。ILE 允许输入锁存信号,高电平有效,输入寄存器的锁存信号由 ILE、$\overline{CS}$、$\overline{WR1}$的逻辑组合产生。当 ILE 为高电平、$\overline{CS}$为低电平、$\overline{WR1}$输入负脉冲时,输入寄存器的锁存信号产生正脉冲。当输入寄存器的锁存信号为高电平时,输入线的状态变化,输入寄存器的锁存信号的负跳变将输入在数据线上的信息送入输入锁存器。

$\overline{WR1}$——写信号 1,低电平有效。当$\overline{WR1}$、$\overline{CS}$、ILE 均有效时,可将数据写入 8 位输入寄存器。

$\overline{WR2}$——写信号 2,低电平有效。当$\overline{WR2}$有效时,在$\overline{XFER}$传送控制信号的作用下,可将锁存在输入寄存器的 8 位数据送到 DAC 寄存器。

$\overline{XFER}$——数据传送信号,低电平有效。当$\overline{WR2}$、$\overline{XFER}$均有效时,则在 DAC 寄存器的锁存信号产生正脉冲,当 DAC 寄存器的锁存信号为高电平时,DAC 寄存器的输出和输入寄存器的状态一致,DAC 寄存器的锁存信号负跳变,将输入寄存器的内容送入 DAC 寄存器。

V_{ref}——基准电源输入端,它与 DAC 内的 R－2R 梯形网络相接,V_{ref}可在$\pm 10V$范围内调节。

DI0~DI7——8 位数字量输入端,ID7 为最高位,DI0 为最低位。

I_{out1}——DAC 的电流输出 1,当 DAC 寄存器各位为 1 时,输出电流为最大。当 DAC 寄存器各位为 0 时,输出电流为 0。

I_{out2}——DAC 的电流输出 2,它使 $I_{out1}+I_{out2}$ 恒为一常数。一般在单极性输出时 I_{out2} 接地,在双极性输出时接运放。

R_{fb}——反馈电阻。在 DAC0832 芯片内有一个反馈电阻,可用作外部运放的分路反馈电阻。

V_{CC}——电源输入线。DGND 为数字地,AGND 为模拟信号地。

从 DAC0832 的时序图 23-3 可以看出,在 CS 有效时,数据更新在 WR 的下降沿锁定。

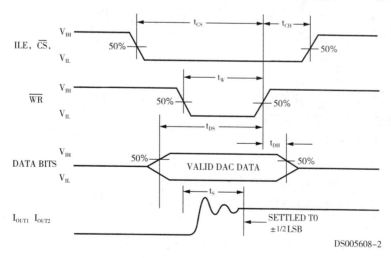

图 23-3 DAC0832 时序图

二、算法设计

由 PC 机发送到单片机串口的值作为数字量初始值,通过按键 K1、K2 实现 DAC 值的加减,按键 K1 实现数字量初始值加 1,按键 K2 实现数字量初始值减 1,使数字量在 20~200

的范围内变化,从而模拟数字信号的输入。

根据 D/A 转换芯片 DAC0832 的工作时序,20ms 进行一次 D/A 转换,可以利用定时器 T0 定时。基本定时时间为 5ms,控制软计数器的累计次数为 4 次,20ms(4×5ms)后定时到时,产生定时器 T0 中断。在定时器 T0 中断服务函数中调用 DAC0832 采样转换函数进行 D/A 采样转换,然后调用计算 D/A 转换值函数把 D/A 转换值转换成相应的 ASCII 码,最后通过 LCD 显示 D/A 转换值函数把 DAC 转换的模拟电压值(0~5V)显示在液晶 LCD1602 上,并从"DA 输出"端子输出 DAC 转换的模拟电压,可用万用表测量。

三、程序设计

1. 主函数设计

主函数主要完成硬件初始化、数据初始化、函数调用等功能。

(1)初始化

首先调用 LCD 初始化函数,并进行数据初始化:按键标志 keybit 和串口成功接收数据标志位 recokbit 为 0,当前 D/A 值 dabl 为 20,软计数器 keytime 为 4。然后设置 LCD 的 DDRAM 地址为 00H,并在 LCD 上显示字符数据"DAC:"。

(2)定时初值计算

定时器 T0 定时时间为 5ms,系统所用的石英晶体振荡频率为 11.0592MHz,因此,1 个机器周期=1/石英频率×12,即为 12/11.0592us,定时器工作方式设置为方式 1,计算初值如下:

$$X=2^{16}-t \times f_{osc}/12=65536-5 \times 10^{-3} \times 11.0592 \times 10^{6}/12=60928=0xee00$$

所以:TH0=0xee,TL0=0x00。

(3)定时器与串口设置

设置寄存器 SCON 的 SM0、SM1 位,定义串口工作方式,设置 SCON=0x50。

选择定时器 T1 作为波特率发生器,设定定时器 T1 的工作方式为方式 2。设置波特率参数为 9600bps,TH1=0xfd,TL1=0xfd。

允许串口接收数据,定义 REN=1。

启动定时/计数器 T1 工作,定义 TR1=1。

设定定时器 T0 的工作方式为方式 1;启动定时器 T0,即 TR0=1。

开放总中断、串口中断和定时器 T0 中断。

设置 TMOD=0x21,IE=0x92。

(4)等待中断

CPU 等待串口中断或定时器 T0 中断的到来。

单片机串口从 PC 机接收到数据后,串口接收中断标志位 RI 置 1,进入串口中断服务函数。

定时器 T0 启动计时后,当定时器 T0 定时 5ms 后,进入定时器 T0 中断服务函数。

主函数设计流程图如图 23-4 所示。

2. 定时器 T0 中断服务函数模块设计

当定时器 T0 定时 5ms 后,进入定时器 T0 中断服务函数。

首先重装定时器 T0 初值,即 TH0=0xee,TL0=0x00。每定时 5ms 一次,软计数器 keytime 值减 1。

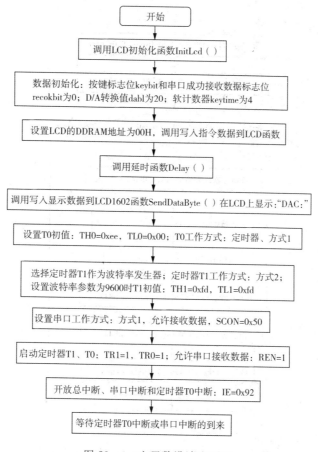

图 23-4 主函数设计流程图

然后判断软计数器 keytime 值是否为 0：

(1)若 keytime 值不为 0，表明 20ms(20ms 采样一次)计时未到，这时中断函数返回主程序，继续计时。

(2)若 keytime 值为 0，表明 20ms(20ms 采样一次)计时已到，重置软计数器 keytime 初值为 4，为下次定时做准备。设置按键输入口 P1 口的值为 0xdf，调用按键扫描函数。接着调用 D/A 转换函数进行 D/A 转换，得到 D/A 转换值 dabl，然后调用计算 D/A 转换值函数计算 dabl 相应的 ASCII 码，再调用 LCD 显示 D/A 转换值函数把数字量输入值 D/A 进行转换后的模拟量显示在液晶 LCD1602 上。

最后 T0 中断函数返回主程序进行下一次 D/A 转换。

定时器 T0 中断服务函数设计流程图如图 23-5 所示。

3.DAC0832 采样转换函数模块设计

根据 D/A 转换芯片 DAC0832 的工作时序，片选信号$\overline{CS}$低电平有效。把芯片 DAC0832 的片选信号$\overline{CS}$置低电平，选中该芯片。

把数字量 D/A 转换值 dabl 送给 P0 口，准备送入芯片 DAC0832 进行 D/A 转换处理。

芯片 DAC0832 写信号$\overline{WR}$置低电平后再过两个时钟周期至高电平，产生一个上升沿信号，进行数据 D/A 转换处理。

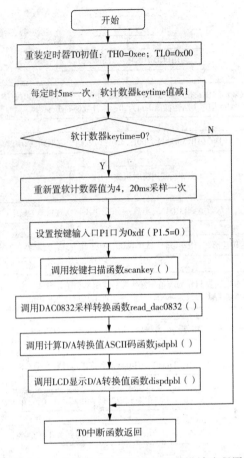

图 23-5　定时器 T0 中断服务函数设计流程图

模拟量 dab1 从 I_{out1}、I_{out2} 引脚输出送到运算放大器进行处理，最后从"DA 输出"端子输出，可用万用表测量到输出电压。

数据转换完毕，把芯片 DAC0832 片选信号 $\overline{CS}$ 置高电平，最后函数返回。

D/A 转换函数设计流程图如图 23-6 所示。

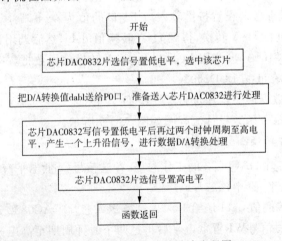

图 23-6　D/A 转换函数设计流程图

4. 计算 D/A 转换值 ASCII 码函数模块设计

要把 DA 转换值 dabl 显示在 LCD 上，需要把它转换为相应的 ASCII 码。

首先计算 D/A 转换值百位数 ASCII 码：将 D/A 转换值 dabl 除以 100 得到的商与 0x30（因为字符数字 0～9 与其相应的 ASCII 码相差 30H）相与。

然后计算 D/A 转换值十位数 ASCII 码：将 D/A 转换值 dabl 除以 100 得到的余数再除以 10，得到的商与 0x30（因为字符数字 0～9 与其相应的 ASCII 码相差 30H）相与。

再计算 D/A 转换值个位数 ASCII 码：将 D/A 转换值 dabl 除以 10 得到的余数与 0x30（因为字符数字 0～9 与其相应的 ASCII 码相差 30H）相与。

最后函数返回。

计算 D/A 转换值 ASCII 码函数设计流程图如图 23-7 所示。

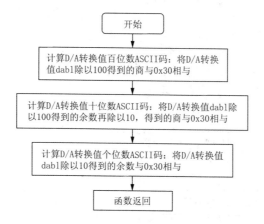

图 23-7 计算 D/A 转换值 ASCII 码函数设计流程图

5. 按键扫描函数模块设计

通过按键 K1、K2 实现 DAC 值 dabl 的加减。程序通过 P1.0、P1.1 的值来判断是否有键按下。

首先把 P1 口所有引脚置为高电平，即 P1=0xff，准备从 P1 口读取数据。

接着判断 K1 键（+键）是否按下：

(1)若 P1.0=0，表明 K1 键（+键）按下，再根据按键标志位 keybit 值判断是否为上次按下键。

若按键标志位 keybit=0，表明本次有键按下，把按键标志位 keybit 置 1，使 DAC 值 dabl 加 1，判断 D/A 值 dabl 是否大于 200，若 D/A 值 dabl 大于 200，重置 D/A 值 dabl 为 20。

若按键标志位 keybit=1，表明本次无键按下，把 P1 口所有引脚置为高电平，准备从 P1 口读取数据。

(2)若 P1.0=1，表明 K1 键（+键）未按下。

然后判断 K2 键（-键）是否按下：

(3)若 P1.1=0，表明 K2 键（-键）按下，再根据按键标志位 keybit 值判断是否为上次按下键。

若按键标志位 keybit=0，表明本次有键按下，把按键标志位 keybit 置 1，使 DAC 值

dabl 减 1,判断 D/A 值 dabl 是否小于 20,若 D/A 值 dabl 小于 20,重置 D/A 值 dabl 为 200。

若按键标志位 keybit＝1,表明本次无键按下,把 P1 口所有引脚置为高电平,准备从 P1 口读取数据。

当 K1 键(＋键 P1.0)、K2 键(－键 P1.1)均未按下时,把按键标志位 keybit 置 0,把 P1 口所有引脚置为高电平,准备从 P1 口读取数据,等待下一次按键。

按键扫描函数设计流程图如图 23-8 所示。

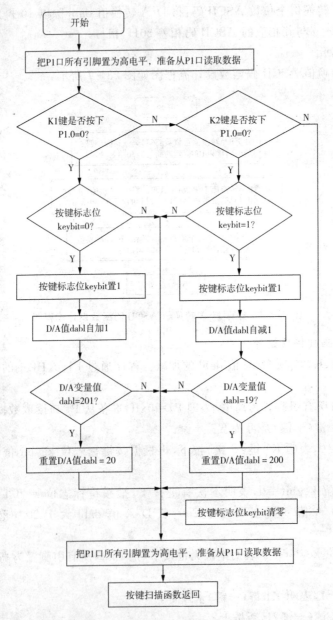

图 23-8 按键扫描函数设计流程图

6. 串口中断服务函数模块设计

当 CPU 检测到串口接收中断标志位 RI 为 1 时,进入串口接收数据中断服务函数,开始从 PC 机串口接收数据。把串口中断标志位 RI 复位置 0,把串口接收缓冲器 SBUF 中的数据送给 DA 转换值变量 dabl 作为 D/A 转换的初始值,同时设置串口成功接收数据标志位 recokbit 为 1,表明串口成功接收数据,最后串口中断函数返回。串口中断服务函数设计流程图如图 23-9 所示。

图 23-9 串口中断服务函数设计流程图

四、C 语言源程序

```
//**********************************************************
//项目名称:D/A 转换
//功能:串口发送一个要输出的电压 KK(00~FF),数/
模转换输出一个当前串口发出的电压;输出的电
//压为 V=5×KK/255。其中 KK 为串口接收到的二进制数
//**********************************************************
#include <reg51.h>         //包含"51 寄存器定义"头文件

//**********************************************************
//LCD1602 信号接口定义
//**********************************************************
sbit E = P2^2;             //LCD 使能信号
sbit RW = P2^1;            //读/写选择信号 R/W:0 为写入数据;1 为读出数据
sbit RS = P2^0;            //数据/命令选择信号 R/S:0 为指令信号;1 为数据信号

;**********************************************************
;功能:D/A 芯片 DAC0832 芯片接口定义
;**********************************************************
sbit DAC_CS = P3^3         //D/A 芯片 DAC0832 的片选信号
sbit DAC_WR = P3^4         //D/A 芯片 DAC0832 的写信号

//**********************************************************
//定义变量:
//dabl—当前 DA 的变量(00 到 255 间),从串口接收
//ledbai—模拟量输出值百位显示值
//ledshi—模拟量输出值十位显示值
//ledge—模拟量输出值个位显示值
//keytime—软计数器
//myflag—定义一个可位寻址的片内数据存储单元(bdata)
//keybit—按键标志位 keybit
//recokbit—串口成功接收数据标志位
//**********************************************************
unsigned char dabl,ledbai,ledshi,ledge,keytime;
unsigned char bdata myflag;//定义一个可位寻址的片内数据存储单元(bdata)
```

```c
sbit keybit = myflag^0;        //按键标志位 keybit
sbitrecokbit = myflag^1;       //串口成功接收数据标志位
```

// **
//功能:函数声明
// **
```c
void rs232(void);                       //串口中断服务函数 rs232()
void time0(void);                       //定时器 T0 中断服务函数 time0()
void scankey(void);                     //按键扫描函数 scankey()
void read_dac0832(void);                // DAC0832 采样转换函数 read_dac0832()
void Delay(unsigned int t);             //延时函数 Delay()
void InitLcd(void);                     //LCD 初始化函数 InitLcd()
void SendCommandByte(unsigned char ch); //写入指令数据到 LCD1602 函数
void SendDataByte(unsigned char ch);    //写入显示数据到 LCD1602 函数
void jsdpbl(void);                      //计算 D/A 转换值 ASCII 码函数 jsdpbl()
void dispdpbl(void);                    //LCD 显示 D/A 转换值函数 dispdpbl()
```

// **
//功能:主函数
// **
```c
void main(void)              //主函数入口地址
{
    InitLcd();               //调用 LCD 初始化函数 InitLcd()
    myflag = 0x00;           //初始化 keybit 和 recokbit 为 0
    dabl = 20;               //初始化当前 DA 值 dabl 为 20
    keytime = 4;             //初始化软计数器值为 4,20ms 扫描一次按键
    SendCommandByte(0x80);   //设置 LCD 的 DDRAM 地址为 00H
    Delay(2);                //调用延时函数
    SendDataByte('D');       //调用写入显示数据到 LCD1602 函数
    SendDataByte('A');       //调用写入显示数据到 LCD1602 函数
    SendDataByte('C');       //调用写入显示数据到 LCD1602 函数
    SendDataByte(':');       //调用写入显示数据到 LCD1602 函数
    TH0 = 0xee;              //定时器 T0 定时 5ms 初值设置:TH0 = 0xee
    TL0 = 0x00;              //定时器 T0 定时 5ms 初值设置:TL0 = 0x00
    //定时器 T0 工作方式设置:定时器、方式 1
    //定时器 T1 作为波特率发生器,工作模式为 2
    TMOD = 0x21;
    TH1 = 0xfd;              //串行通信波特率为 9600 时,T1 初值:TH1 = 0xfd
    TL1 = 0xfd;              //串行通信波特率为 9600 时,T1 初值:TL1 = 0xfd
    SCON = 0x50;             //设置串口工作方式:串口方式 1,允许接收数据
    TR1 = 1;                 //启动定时器 T1
    REN = 1;                 //允许串口接收数据
    TR0 = 1;                 //启动定时器 T0
```

```c
    IE = 0x92;                    //开放总中断、串口中断和定时器 T0 中断
    while(1)                      //等待串口中断或定时器 T0 中断的到来
    { }
}

//*********************************************************************
//功能:串口中断函数 rs232()
//*********************************************************************
void rs232(void) interrupt 4      //串口中断函数 rs232()
{
    RI = 0;                       //把串口接收中断标志位 RI 复位置 0
    dabl = SBUF;                  //把单片机从 PC 串口接收的数据 SBUF 送给 D/A 转换值 dabl
    recokbit = 1;                 //串口成功接收数据标志位 recokbit 置 1
}


//*********************************************************************
//功能:定时器 T0 中断服务函数 time0( )
//*********************************************************************
void time0(void) interrupt 1      //定时器 T0 中断服务函数
{
    TH0 = 0xee;                   //重新设置定时器 T0 初值:TH0 = 0xee
    TL0 = 0x00;                   //重新设置定时器 T0 初值:TL0 = 0x00
    keytime - -;                  //每定时 5ms 一次,软计数器 keytime 值减 1
    if(keytime = = 0)             //判断 20ms 定时时间是否到达,如到达,开始 D/A 采样
    {
        keytime = 4;              //重新置软计数器值为 4,20ms 采样一次
        P1 = 0xdf;                //设置按键输入口 P1 = 0xdf(P1.5 = 0)
        scankey();                //调用按键扫描函数 scankey()
        read_dac0832();           //调用 DAC0832 采样转换函数 read_dac0832()
        jsdpbl();                 //调用计算 D/A 转换值 ASCII 码函数 jsdpbl()
        dispdpbl();               //调用 LCD 显示 D/A 转换值函数 dispdpbl()
    }
}


//*********************************************************************
//功能:;按键扫描函数 scankey( )
//*********************************************************************
void scankey(void)
{
    P1 = 0xff;                    //把 P1 口所有引脚置为高电平,准备从 P1 口读取数据

    //情况 1:K1 键( + 键 P1.0)按下
    if(P1.0 = = 0)                //判断 K1 键( + 键 P1.0)是否按下
```

```c
    {
            if(keybit==0)              //根据按键标志位 keybit 判断是否为上次按下键
            {   keybit=1;              //若 keybit=0,表明本次有键按下,并把 keybit 置 1
                dabl++;                //D/A 值 dabl 自加 1
                if(dabl==201)          //判断 D/A 值 dabl 是否大于 200
                { dabl=20; }           //D/A 值 dabl 大于 200,重置 D/A 值 dabl 为 20
            }
    }

    //情况 2:K2 键(-键 P1.1)按下
    else if(P1.1==0)                   //判断 K2 键(-键 P1.1)是否按下
    {
            if(keybit==0)              //根据按键标志位 keybit 判断是否为上次按下键
            {
                keybit=1;              //若 keybit=0,表明本次有键按下,并把 keybit 置 1
                dabl--;                //D/A 值 dabl 自减 1
                if(dabl==19)           //判断 D/A 值 dabl 是否小于 20
                { dabl=200; }          //D/A 值 dabl 小于 20,则重置 D/A 值 dabl 为 200
            }
    }

    //情况 3:K1 键(+键 P1.0)、K2 键(-键 P1.1)均未按下
    else { keybit=0; }  //K1、K2 键均未按下,把 keybit 置 0,等待下一次按键

    P1=0xff;                           //把 P1 口所有引脚置为高电平,准备从 P1 口读取数据
}

//*****************************************************************************
//功能:D/A 采样转换函数 read_dac0832( )
//*****************************************************************************
void read_dac0832(void)                //DAC0832 采样转换函数
{
    DAC_CS=0;                          //芯片 DAC0832 片选信号 DAC_CS 置低电平(低电平有效)
    P0=dabl;                           //把当前 DA 值 dabl 送给 P0,准备送入 DAC0832 进行 D/A 处理
    DAC_WR=0;                          //芯片 DAC0832 写信号 DAC_WR 置低电平
    Delay(1);                          //延时
    DAC_WR=1;                          //芯片 DAC0832 写信号 DAC_WR 置高电平,产生一个上升沿信号
    DAC_CS=1;                          //芯片 DAC0832 片选信号 DAC_CS 置高电平(高电平无效)
}

//*****************************************************************************
//功能:延时 40us 函数
//*****************************************************************************
```

任务 23 D/A 转换

```c
void Delay(unsigned int t)        // delay 40us
{
    for(;t! = 0;t - - );
}


// ****************************************************************************
//功能:LCD1602 初始化函数 InitLcd( )
// ****************************************************************************
void InitLcd( )
{
    SendCommandByte(0x30);        //显示模式设置(不测试忙信号)共三次
    SendCommandByte(0x30);        //显示模式设置(不测试忙信号)共三次
    SendCommandByte(0x30);        //显示模式设置(不测试忙信号)共三次
    SendCommandByte(0x38);        //设置工作方式:8 位数据口,2 行显示,5*7 点阵
    SendCommandByte(0x08);        //设置显示状态:显示关
    SendCommandByte(0x01);        //清屏
    SendCommandByte(0x06);        //设置输入方式:数据读写后 AC 自增 1,画面不动
    SendCommandByte(0x0c);        //设置显示状态:显示开
}


// ****************************************************************************
//功能:写入指令数据到 LCD1602 函数 SendCommandByte()
//形参:ch
// ****************************************************************************
void SendCommandByte(unsigned char ch)
{
    RS = 0;                       //数据/命令选择信号 R/S:0 为指令信号;1 为数据信号
    RW = 0;                       //读/写选择信号 R/W:0 为写入数据;1 为读出数据
    P0 = ch;                      //把指令数据 ch 送至 P0 口(LCD 数据线 DB7~DB0)
    E = 1;                        //LCD 使能信号 E 置高电平
    Delay(1);                     //延时 0.40us
    E = 0;                        //LCD 使能信号 E 置低电平
    Delay(100);                   //delay40us
}


// ****************************************************************************
//功能:写入显示数据到 LCD1602 函数 SendDataByte()
//形参:ch
// ****************************************************************************
void SendDataByte(unsigned char ch)
{
    RS = 1;                       //数据/命令选择信号 R/S:0 为指令信号;1 为数据信号
    RW = 0;                       //读/写选择信号 R/W:0 为写入数据;1 为读出数据
```

```
    P0 = ch;                    //把显示数据 ch 送至 P0 口(LCD 数据线 DB7~DB0)
    E = 1;                      //LCD 使能信号 E 置高电平
    Delay(1);                   //延时 0.40us
    E = 0;                      //LCD 使能信号 E 置低电平
    Delay(100);                 //delay 40us
}
```

// **
//功能:计算 D/A 转换值 ASCII 码函数 jsdpbl()
// **
```
void jsdpbl(void)
{
    //计算 D/A 转换值百位,并与 0x30 相与,得到 D/A 转换值百位的 ASCII 码
    ledbai = (dabl/100)|0x30;
    //计算 D/A 转换值十位,并与 0x30 相与,得到 D/A 转换值十位的 ASCII 码
    ledshi = ((dabl%100)/10)|0x30;
    //计算 D/A 转换值个位,并与 0x30 相与,得到 D/A 转换值个位的 ASCII 码
    ledge = (dabl%10)|0x30;
}
```

// **
//功能:LCD 显示 D/A 转换值函数 dispdpbl()
// **
```
void dispdpbl(void)             //LCD 显示 D/A 转换值函数
{
    SendCommandByte(0x85);      //设置 LCD 的 DDRAM 地址为 05H
    SendDataByte(ledbai);       //LCD 显示 D/A 转换值百位
    SendDataByte(ledshi);       //LCD 显示 D/A 转换值十位
    SendDataByte(ledge);        //LCD 显示 D/A 转换值个位
}
```

五、汇编语言源程序

```
;************************************************************************
;项目名称:D/A 转换
;功能:串口发送一个要输出的电压 KK(00~FF),数/模转换输出一个当前串口发出的电压;
;输出的电压为 V=5×KK/255。其中 KK 为串口接收到的二进制数
;************************************************************************


;************************************************************************
;功能:LCD1602 与单片机信号接口定义
;************************************************************************
E       BIT     P2.2            ;LCD 使能信号
RW      BIT     P2.1            ;读/写选择信号 R/W:0 为写入数据;1 为读出数据
```

```
RS          BIT     P2.0            ;数据/命令选择信号 R/S:0 为指令;1 为数据
LCDPORT     EQU     P0              ;LCD1602 数据线 DB7~DB0
CMD_BYTE    EQU     2EH             ;写指令数据入口参数
DAT_BYTE    EQU     2FH             ;写显示数据入口参数

;**********************************************************************
;功能:DAC0832 的引脚定义
;**********************************************************************
DAC_WR      BIT     P3.4            ;D/A 芯片 DAC0832 的写信号
DAC_CS      BIT     P3.3            ;D/A 芯片 DAC0832 的片选信号

;**********************************************************************
;功能:
;**********************************************************************
DABL        EQU     30H             ;当前 DA 的变量(00 到 255 间),从串口接收
LEDBAI      EQU     31H             ;显示的百位
LEDSHI      EQU     32H             ;显示的十位
LEDGE       EQU     33H             ;显示的个位
KEYTIME     EQU     34H             ;软计数器

KEYBIT      BIT     00H             ;按键标志位
RECOKBIT    BIT     01H             ;串口成功接收数据标志位

ORG         0000H
AJMP        MAIN

;**********************************************************************
;功能:定时器 T0 中断服务入口地址
;**********************************************************************
            ORG     000BH           ;T0 中断服务入口地址
            AJMP    TIME0_1         ;跳转到 T0 中断服务子程序 TIME0_1

;**********************************************************************
;功能:串口接收数据中断服务子程序 RS232
;**********************************************************************
            ORG     0023H           ;串口中断服务入口地址 0023H
RS232:      CLR     RI
            MOV     A,SBUF          ;单片机从 PC 串口接收数据
            MOV     DABL,A          ;PC 串口发送的数据送给 D/A 转换值
            SETB    RECOKBIT        ;设置串口成功接收数据标志位 RECOKBIT 为 1
            RETI                    ;串口接收数据中断服务子程序返回

;**********************************************************************
```

;功能:主程序
;**
            ORG   0030H
MAIN: CLR   KEYBIT   ;初始化按键标志位 KEYBIT 为 0
            CLR   RECOKBIT   ;初始化串口成功接收数据标志位 RECOKBIT 为 0
            MOV   DABL,#20   ;初始化 D/A 转换值 DABL 为 20
            MOV   KEYTIME,#04H   ;软计数器 KEYTIME 为 4
            LCALL   INITLCD   ;调用 LCD 初始化子程序
            MOV   CMD_BYTE,#80H   ;设置 DDRAM 的地址
            LCALL   WRITE_CMD   ;调用写入指令数据到 LCD1602 子程序
            LCALL   DELAY0   ;调用延时子程序

            MOV   DAT_BYTE,#"D"
            LCALL   WRITE_DAT   ;调用写入显示数据到 LCD1602 子程序
            MOV   DAT_BYTE,#"A"
            LCALL   WRITE_DAT   ;调用写入显示数据到 LCD1602 子程序
            MOV   DAT_BYTE,#"C"
            LCALL   WRITE_DAT   ;调用写入显示数据到 LCD1602 子程序
            MOV   DAT_BYTE,#":"
            LCALL   WRITE_DAT   ;调用写入显示数据到 LCD1602 子程序

            MOV   TH0,#0EEH   ;定时器 T0 定时 5ms,D/A 转换时间
            MOV   TL0,#00H
            MOV   TMOD,#21H
            MOV   TH1,#0FDH   ;T1 为波特率发生器,波特率 9600bps
            MOV   TL1,#0FDH
            MOV   SCON,#50H   ;设置串口工作方式
            SETB   TR1   ;启动定时器 T1
            SETB   REN   ;允许串行口接收数据
            SETB   TR0   ;启动定时器 T0
            MOV   IE,#92H   ;开放总中断、定时器 T0 中断、串口中断
            SJMP $;等待串口中断、定时器 T0 中断

;**
;功能:定时器 T0 中断服务子程序 TIME0_1
;**
            ORG   0100H
TIME0_1:   MOV   TH0,#0EEH
            MOV   TL0,#00H
            DJNZ   KEYTIME,TIME0_RE
            MOV   KEYTIME,#04H
            MOV   P1,#0DFH
            LCALL   KEYSCAN

```
            LCALL      DAC0832           ;调用 D/A 转换子程序 DAC0832
            LCALL      JSDPBL            ;调用计算 D/A 转换值 ASCII 码子程序
            LCALL      DISPDPBL          ;调用 LCD 显示 D/A 转换值子程序
TIME0_RE:   RETI                         ;中断子程序返回

;*******************************************************************************
;功能:;按键扫描子程序 KEYSCAN
;*******************************************************************************
KEYSCAN:    MOV        P1,#0FFH
NEXT_UP:    JB         P1.0,NEXT_DN      ;+ 键
            JB         KEYBIT,SCAN_RE
            SETB       KEYBIT
            INC        DABL
            MOV        A,DABL
            CJNE       A,#201,SCAN_RE    ;若 DABL 大于 200 返回 20
            MOV        DABL,#20
            AJMP       SCAN_RE
NEXT_DN:    JB         P1.1,NEXT_NC      ;- 键
            JB         KEYBIT,SCAN_RE
            SETB       KEYBIT
            DEC        DABL
            MOV        A,DABL
            CJNE       A,#19,SCAN_RE     ;若 DABL 小于 20 返回 200
            MOV        DABL,#200
            AJMP       SCAN_RE
NEXT_NC:    CLR        KEYBIT            ;清标志用于等待下一次按键
SCAN_RE:    MOV        P1,#0FFH
            RET                          ;子程序返回

;*******************************************************************************
;功能:;DAC0832 采样转换子程序 DAC0832
;*******************************************************************************
DAC0832:    CLR        DAC_CS
            MOV        A,DABL
            MOV        P0,A
            CLR        DAC_WR
            NOP
            NOP
            SETB       DAC_WR
            SETB       DAC_CS
            RET                          ;子程序返回

;*******************************************************************************
```

;功能:LCD1602要用到的一些子程序
;**
;**
;功能:写入指令数据到LCD1602子程序 WRITE_CMD（入口参数 CMD_BYTE）
;**
WRITE_CMD: CLR   RS
           CLR   RW
           MOV   A,CMD_BYTE
           MOV   LCDPORT,A
           SETB   E
           NOP
           NOP
           CLR   E
           LCALL   DELAY0
           RET   ;写入指令数据到LCD1602子程序返回

;**
;功能:写入显示数据到LCD1602子程序 WRITE_DAT（入口参数 DAT_BYTE）
;**
WRITE_DAT: SETB   RS
           CLR   RW
           MOV   A,DAT_BYTE
           MOV   LCDPORT,A
           SETB   E
           NOP
           NOP
           CLR   E
           LCALL   DELAY0
           RET   ;写入显示数据到LCD1602子程序返回

;**
;功能:LCD初始化子程序 INITLCD
;**
INITLCD: MOV   CMD_BYTE,#30H
           LCALL   WRITE_CMD   ;调用写入指令数据到LCD1602函数
           MOV   CMD_BYTE,#30H
           LCALL   WRITE_CMD   ;调用写入指令数据到LCD1602子程序
           MOV   CMD_BYTE,#30H
           LCALL   WRITE_CMD   ;调用写入指令数据到LCD1602子程序
           MOV   CMD_BYTE,#38H   ;设定工作方式
           LCALL   WRITE_CMD   ;调用写入指令数据到LCD1602子程序
           MOV   CMD_BYTE,#0CH   ;显示状态设置
           LCALL   WRITE_CMD   ;调用写入指令数据到LCD1602子程序

```
            MOV      CMD_BYTE,#01H       ;清屏
            LCALL    WRITE_CMD           ;调用写入指令数据到LCD1602子程序
            MOV      CMD_BYTE,#06H       ;输入方式设置
            LCALL    WRITE_CMD           ;调用写入指令数据到LCD1602子程序
            RET                          ;LCD初始化子程序返回

;***************************************************************
;功能:延时子程序
;***************************************************************
DELAY0:     MOV      R5,#0A0H
DELAY1:     NOP
            DJNZ     R5,DELAY1
            RET                          ;子程序返回


;***************************************************************
;功能:计算D/A转换值ASCII码子程序JSDPBL
;***************************************************************
JSDPBL:     MOV      A,DABL              ;取得A/D转换值DABL
            MOV      B,#64H              ;除以100,得到百位
            DIV      AB
            ORL      A,#30H              ;加30H,变换成相应的ASCII
            MOV      LEDBAI,A            ;送到LCD显示

            MOV      A,B
            MOV      B,#0AH              ;除以10,得到十位
            DIV      AB
            ORL      A,#30H              ;加30H,变换成相应的ASCII
            MOV      LEDSHI,A            ;送到LCD显示

            MOV      A,B                 ;余数为个位
            ORL      A,#30H              ;加30H,变换成相应的ASCII
            MOV      LEDGE,A             ;送到LCD显示

            RET                          ;计算D/A转换值子程序返回

;***************************************************************
;功能:LCD显示D/A转换值子程序DISPDPBL
;***************************************************************
DISPDPBL:   MOV      CMD_BYTE,#85H       ;设置LCD的DDRAM的地址为05H
            LCALL    WRITE_CMD           ;调用写入指令数据到LCD1602函数
            MOV      DAT_BYTE,LEDBAI     ;D/A转换值百位显示
            LCALL    WRITE_DAT           ;调用写入显示数据到LCD1602子程序
            MOV      DAT_BYTE,LEDSHI     ;D/A转换值十位显示
```

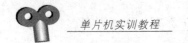

```
        LCALL    WRITE_DAT           ;调用写入显示数据到LCD1602子程序
        MOV      DAT_BYTE,LEDGE      ;D/A转换值个位显示
        LCALL    WRITE_DAT           ;调用写入显示数据到LCD1602子程序
        RET                          ;LCD显示D/A转换值子程序返回

        END                          ;程序结束
```

实训考核

本课程改革传统的闭卷或开卷考核,而采用过程考核为主的多元化考核方式,考核分为理论考核、职业道德考核和技能考核三部分,各部分所占比例见表23-1、表23-2、表23-3。

表23-1 理论考核和职业素质考核形式及所占比例

序号	名称			比例	得分
一	理论考核	过程作业文件	个人自评	10%	
			组内互评	20%	
			小组互评	30%	
			老师评定	20%	
		课堂提问、解答		10%	
		项目汇报		10%	
		小计		100%	
二	职业素质	职业道德,工作作风		40%	
		小组沟通协作能力		40%	
		创新能力		20%	
		小计		100%	

表23-2 技能考核内容及比例

姓名		班级		小组		总得分	
序号	考核项目	考核内容及要求	配分	评分标准	考核环节	得分	
1	①安全文明生产 ②安全操作规范	着装规范	20%	现场考评	实施		
		安全用电					
		布线规范、合理					
		工具摆放整齐					
		工具及仪器仪表使用规范、摆放整齐					
		任务完成后,进行场地整理,保持场地清洁、有序					

（续表）

序号	考核项目	考核内容及要求	配分	评分标准	考核环节	得分
2	实训态度	不迟到、早退、旷课	10%	现场考评	六步	
		实训过程认真负责				
		组内主动沟通、协作，小组间互助				
3	系统方案制定	工作流程正确合理	10%	现场考评	计划决策	
		方案合理				
		选用指令是否合理				
		电路图正确				
	编程能力	独立完成程序	10%	现场考评	决策	
		程序简单、可靠				
4	操作能力	正确输入程序并进行程序调试	20%	现场考评	实施	
		根据电路图正确接线				
		根据系统功能进行正确操作演示				
5	工艺	接线美观	10%	现场考评	实施	
		线路工作可靠				
6	实践效果	系统工作可靠	10%	现场考评	检查	
		满足工作要求				
		创新				
		按规定的时间完成项目				
7	汇报总结	工作总结，PPT汇报	5%	现场考评	评估	
		填写自我检查表及反馈表				
8	技术文件制作整理	技术文件制作整理能力	5%	现场考评	评估	
	合计		100%			

表23-3 各部分考核占课程考核的比例

考核项目	理论考核	技能考核	职业素质考核	合计
比例	30%	50%	20%	100%
分值	30	50	20	100
实际得分				

任务 24　I²C 总线存储器读写

实训任务

读取 I²C 总线存储器 24C02 里的内容，并且在 LCD1602 上显示出当前读出的地址和当前地址的内容。

实训设备

1. 设备

PC 机（安装 wave 编程软件、Keil C51 软件）、单片机实验板。

2. 工具及材料

工作对象：电工电子工具、电子元器件和辅助材料、仿真器、编程器。

工作工具：单片机控制电路原理图、实训指导书、项目任务单、工作记录单、项目检查单、各种电工仪表、常用电工工具和拆装工具、量具、相关电子手册。

硬件设计

主控模块采用 ATMEL 公司生产的 AT89S52 单片机，LCD 显示模块选用 1602 字符型 LCD 模块，I²C 总线存储器选用 24C02，24C02 与单片机的接口电路如图 24-1 所示。

软件设计

一、I²C 总线介绍

1. I²C 总线的特性

在现代电子产品开发过程中，为了简化系统，提高系统的可靠性，缩短产品开发周期，增加硬件构成的灵活性，继而推出了一种高效、可靠、方便的 I²C 串行总线。在单片机应用系统中推广 I²C 总线后大大改变单片机应用系统结构、性能，简化了结构。二线制的 I²C 串行总线使得各电路单元之间只需要简单的 2 线连接，而且总线接口都已集成在器件中。不需另加总线接口电路，这样减少电路板面积。提高了可靠性，降低了成本。并且可实现电路系统的模块化、标准化设计。在 I²C 总线上各单元电路相互之间没有其他连线，用户常用的单元电路基本上与系统电路无关，故极易形成用户自己的标准化、模块化设计。因为这些，使得支持 I²C 总线的器件大量涌现，I²C 得到了广泛的应用。

本节就讨论基于 I²C 总线的 E2PROM 的应用。

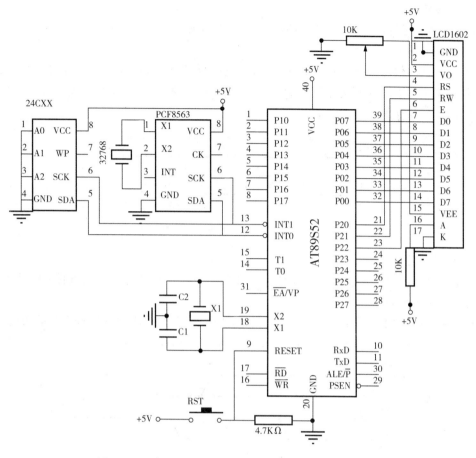

图 24-1 I²C 总线存储器 24C02 与单片机的接口电路图

2. I²C 总线规范

I²C 总线通过两根线——串行数据（SDA）和串行时钟（SCK）线连接到总线上的任何一个器件，每个器件都应有一个唯一的地址，而且都可以作为一个发送器或接收器。此外，器件在执行数据传输时也可以被看作是主机或从机。

发送器：本次传送中发送数据（不包括地址和命令）到总线的器件。

接收器：本次传送中从总线接收数据（不包括地址和命令）的器件。

主机：初始化发送、产生时钟信号和终止发送的器件，它可以是发送器或接收器。主机通常是微控制器。

从机：被主机寻址的器件，它可以是发送器或接收器。

I²C 总线是一个多主机的总线，也即可以连接多于一个能控制总线的器件到总线。当两个以上能控制总线的器件同时发动传输时，只能有一个器件能真正控制总线而成为主机，并使报文不被破坏，这个过程叫做仲裁。与此同时，能使多个能控制总线的器件产生时钟信号的同步。

SDA 和 SCK 都是双向线路，连接到总线的器件的输出级必须是漏极开路或集电极开路，都通过一个电流源或上拉电阻连接到正的电源电压，这样才能够实现线与功能。当总线空闲时，这两条线路都是高电平。在标准模式下，数据传输的速度为 0～100kbit/s。

3. 位传输

I²C 总线上每传输一个数据位必须产生一个时钟脉冲。

(1)数据的有效性

SDA 线上的数据必须在时钟线 SCK 的高电平周期保持稳定,数据线的电平状态只有在 SCK 线的时钟信号是低电平时才能改变,如图 24-2 所示。在标准模式下,高低电平宽度必须不小于 $4.7\mu S$。

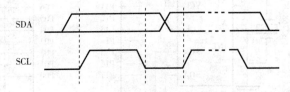

图 24-2 I²C 总线控制的位传输

(2)起始和停止条件

在 I²C 总线中,唯一违反上述数据有效性的是起始(S)和停止(P)条件,如图 24-3 所示。

起始条件(重复起始条件):在 SCK 线是高电平时,SDA 线从高电平向低电平切换。

停止条件:在 SCK 线是高电平时,SDA 线由低电平向高电平切换。

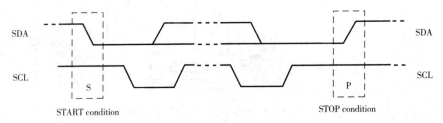

图 24-3 起始位和停止条件

起始和停止条件一般由主机产生。起始条件作为一次传送的开始,在起始条件后总线被认为处于忙的状态。停止条件作为一次传送的结束,在停止条件的某段时间后,总线被认为再次处于空闲状态。重复起始条件既作为上次传送的结束,也作为下次传送的开始。

4. 数据传输

(1)字节格式

发送到 SDA 线上的每个字节必须是 8 位,每次传输可以发送的字节数量不受限制。每个字节后必须跟一个应答位。首先传输的是数据的最高位(MSB)如图 24-4 所示。

(2)应答

相应的应答时钟脉冲由从机产生。在应答的时钟脉冲期间,发送器释放 SDA 线(高)。在应答的时钟脉冲期间,接收器必须将 SDA 线拉低,使它在这个时钟脉冲的高电平期间保持稳定的低电平。如图 24-4 中时钟信号 SCK 的第 9 位。

一般说来,被寻址匹配的从机或可继续接收下一字节的接收器将产生一个应答。若作为发送器的主机在发送完一个字节后,没有收到应答位(或收到一个非应答位),或作为接收器的主机没有发送应答位(或发送一个非应答位),那么主机必须产生一个停止条件或重复起始条件来结束本次传输。

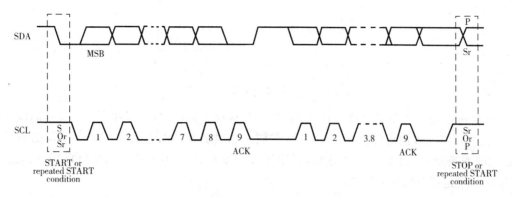

图 24-4 I²C 总线的数据传输

若从机-接收器不能接收更多的数据字节,将不产生这个应答位;主机-接收器在接收完最后一个字节后不产生应答,通知从机-发送器数据结束。

5．仲裁与时钟发生

(1)同步

时钟同步通过各个能产生时钟的器件线与连接到 SCK 线上来实现的,上述的各个器件可能都有自己独立的时钟,各个时钟信号的频率、周期、相位和占空比可能都不相同,由于"线与"的结果,在 SCK 线上产生的实际时钟的低电平宽度由低电平持续时间最长的器件决定,而高电平宽度由高电平持续时间最短的器件决定。

(2)仲裁

当总线空闲时,多个主机同时启动传输,可能会有不止一个主机检测到满足起始条件,而同时获得主机权,这样就要仲裁。当 SCK 线是高电平时,仲裁在 SDA 线发生,当其他主机发送低电平时,发送高电平的主机将丢失仲裁,因为总线上的电平与它自己的电平不同。

仲裁可以持续多位,它的第一个阶段是比较地址位,如果每个主机都尝试寻址相同的器件,仲裁会继续比较数据位,或者比较响应位。因为 I²C 总线的地址和数据信息由赢得仲裁的主机决定,在仲裁过程中不会丢失信息。

(3)用时钟同步机制作为握手

器件可以快速接收数据字节,但可能需要更多时间保存接收到的字节或准备一个要发送的字节。此时,这个器件可以使 SCK 线保持低电平,迫使与之交换数据的器件进入等待状态,直到准备好下一字节的发送或接收。

6．传输协议

(1)寻址字节

主机产生起始条件后,发送的第一个字节为寻址字节,该字节的头 7 位(高 7 位)为从机地址,最低位(LSB)决定了报文的方向,"0"表示主机写信息到从机,"1"表示主机读从机中的信息,如图 24-5 所示。当发送了一个地址后,系统中的每个器件都将头 7 位与它自己的地址比较。如果一样,器件会应答主机的寻址,至于是从机-接收器还是从机-发送器都由 R/W 位决定。

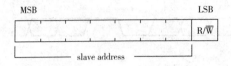

图 24-5 起始条件后的第一个字节

从机地址由一个固定的和一个可编程的部分构成。例如,某些器件有 4 个固定的位(高 4 位)和 3 个可编程的地址位(低 3 位),那么同一总线上共可以连接 8(23)个相同的器件。I²C 总线委员会协调 I²C 地址的分配,保留了 2 组 8 位地址(0000XXX 和 1111XXX),这 2 组地址的用途可查阅有关资料。

(2)传输格式

主机产生起始条件后,发送一个寻址字节,收到应答后跟着就是数据传输,数据传输一般由主机产生的停止位终止。但是,如果主机仍希望在总线上通讯,它可以产生重复起始条件(Sr)和寻址另一个从机,而不是首先产生一个停止条件。在这种传输中,可能有不同的读/写格式结合。可能的数据传输格式有:主机-发送器发送数据到从机-接收器。如图 24-6 所示,寻址字节的"R/W"位为 0,数据传输的方向不改变。

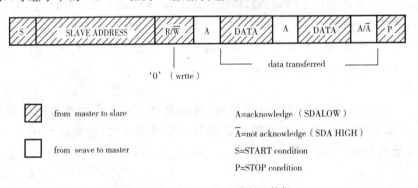

图 24-6 主机-发送器发送数据

寻址字节后,主机-接收器立即读从机-发送器中的数据。如图 24-7 所示,寻址字节的 R/W 位为 1。在第一次从机产生的响应时。主机-发送器变成主机-接收器,从机-接收器变成从机-发送器。之后,数据由从机发送,主机接收,每个应答由主机产生。时钟信号 CLK 仍由主机产生。

若主机要终止本次传输,则发送一个非应答信号(A),接着主机产生停止条件。

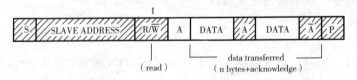

图 24-7 寻址字节后,主机-接收器立即读数据

复合格式,如图 24-8 所示。传输改变方向的时候,起始条件和从机地址都会被重复。但 R/W 位取反。如果主机-接收器发送一个重复起始条件,它之前应该要发送一个非应答信号(A)。

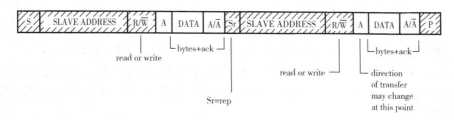

图 24-8 复合格式

7. E2PROM 的应用

(1) 24C02 的应用

电可擦除可编程只读存储器 E2PROM 可分为并行和串行两大类。并行 E2PROM 在读写数据是通过 8 位数据总线传输,而串行 E2PROM 的数据是一位一位的传输。虽然与并行 E2PROM 相比,串行传输数据较慢,但它体积小、专用 I/O 口少、低廉、电路简单等优点,因此广泛用于智能仪器、仪表设备中。

开发板中,提供 ATMEL 公司出品的 24C02。串行 E2PROM 一般具有两种写入方式,一种是字节写入方式,还有另一种页写入方式。允许在一个写周期内同时对 1 个字节到一页的若干字节的编程写入,1 页的大小取决于芯片内页寄存器的大小。

(2) 管脚描述

24CXX 系列 E2PROM 提供标准的 8 脚 DIP 封装和 8 脚表面安装的 SOIC 封装。24C02、24C128、24C256 管脚排列分别如图 24-9 所示。

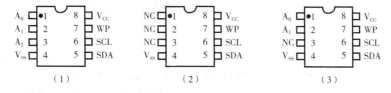

图 24-9 管脚排列图

① SCK:串行时钟。这是一个输入管脚,用于产生器件所有数据发送或接收的时钟。

② SDA:串行数据/地址。这是一个双向传输端,用于传送地址和所有数据的发送或接收。它是一个漏极开路端,因此要求接一个上拉电阻到 V_{cc} 端(典型值为:100KHz 时为 10K,400KHz 时为 1K)。对于一般的数据传输,仅在 SCK 为低期间 SDA 才允许变化。在 SCK 为高期间变化,留给指示 START(开始)和 STOP(停止)条件。

③ A0、A1、A2:器件地址输入端。这些输入端用于多个器件级联时设置器件地址,当这些脚悬空时默认值为 0(开发板上 24C02 的三个脚接地)。

④ WP:写保护。如果 WP 管脚连接到 V_{cc},所有的内容都被写保护(只能读)。当 WP 管脚连接到 V_{ss} 或悬空,允许器件进行正常的读/写操作。

8. 串行 24C02 芯片的寻址

(1) 从器件地址位

主器件通过发送一个起始信号启动发送过程,然后发送它所要寻址的从器件的地址。8 位从器件地址的高 4 位 D7~D4 固定为 1010,接下来的 3 位 D3~D1(A2、A1、A0)为器件的片选地址位或作为存储器页地址选择位,用来定义哪个器件以及器件的哪个部分被主器件访问,最多可以连 8 个 24C02 到同一总线上,这些位(单片机的输出数据)必须与 24C02 的输入脚 A2、A1、A0 相对应。1 个 24C02 可单独被系统寻址。从器件 8 位地址的最低位 D0,作为读写控制位。"1"表示对从器件进行读操作,"0"表示对从器件进行写操作。在主器件发送起始信号和从器件地址字节后,24C02 监视总线并当其地址与发送的从地址相符时响应一个应答信号(通过 SDA 线)。24C02 再根据读写控制位 R/W 的状态进行读或写操作。

(2)应答信号

I²C 总线数据传送时,每成功地传送一个字节数据后,接收器都必须产生一个应答信号。应答的器件在第 9 个时钟周期时将 SDA 线拉低,表示其已收到一个 8 位数据。24C02 在接收到起始信号和从器件地址之后响应一个应答信号,如果器件已选择了写操作,则在每接收一个 8 位字节之后响应一个应答信号。当 24C02 工作于读模式时,在发送一个 8 位数据后释放 SDA 线并监视一个应答信号,一旦接收到应答信号,24C02 继续发送数据,如主器件没有发送应答信号,器件停止传送数据并等待一个停止信号,主器件发一个停止信号给 24C02 使其进入备用电源模式并使器件处于已知的状态。

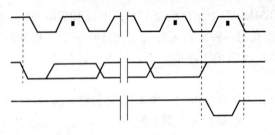

图 24-10 应答时序

9. 写操作方式

(1)字节写

如图 24-11 所示为 24C02 字节写时序图。在字节写模式下,主器件发送起始命令和从器件地址信息(R/W 位置 0)给从器件,主器件在收到从器件产生应答信号后,主器件发送 1 个 8 位字节地址写入 24C02 的地址指针,主器件在收到从器件的另一个应答信号后,再发送数据到被寻址的存储单元。24C02 再次应答,并在主器件产生停止信号后开始内部数据的擦写,在内部擦写过程中,24C02 不再应答主器件的任何请求。

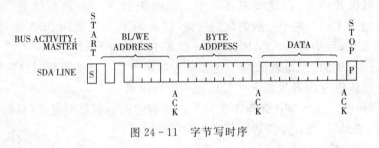

图 24-11 字节写时序

(2)页写

如图 24-12 所示为 24C02 页写时序图,页写模式下 24C02 可一次写入 16 个字节数据。页写操作的启动和字节写一样,不同点在于传送了一字节数据后并不产生停止信号。主器件被允许发送 P=15。每发送一个字节数据 24C02 产生一应答位,且内部低 3 位地址加 1,

高位不变。如果在发送停止信号之前主器件发送超过P+1个字节,地址计数器将自动翻转,先前写入的数据被覆盖。接收到P+1字节数据和主器件发送的停止信号后,24C02启动内部写周期将数据写到数据区。所有接收的数据在一个写周期内写入24C02。

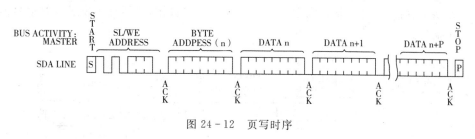

图 24-12 页写时序

(3)应答查询

可以利用内部写周期时禁止数据输入这一特性。一旦主器件发送停止位指示主器件操作结束时,24C02启动内部写周期,应答查询立即启动,包括发送一个起始信号和进行写操作的从器件地址。如果24C02正在进行内部写操作,不会发送应答信号。如果24C02已经完成了内部自写周期。将发送一个应答信号,主器件可以继续进行下一次读写操作。

(4)写保护

写保护操作特性可使用户避免由于不当操作而造成对存储区域内部数据的改写,当WP管脚接高时,整个寄存器区全部被保护起来而变为只可读取。24C02可以接收从器件地址和字节地址,但是装置在接收到第一个数据字节后不发送应答信号从而避免寄存器区域被编程改写。

10. 读操作方式

对24C02读操作的初始化方式和写操作时一样,仅把R/W位置为1,有3种不同的读操作方式:读当前地址内容、读随机地址内容、读顺序地址内容。

(1)立即地址读取

如图24-13所示为24C02立即地址读时序图。24C02的地址计数器内容为最后操作字节的地址加1。也就是说,如果上次读/写的操作地址为N,则立即读的地址从地址N+1开始。如果N=E(24C02,E=255),则计数器将翻转到0且继续输出数据。24C02接收到从器件地址信号后(R/W位置1),它首先发送一个应答信号,然后发送一个8位字节数据。主器件不需发送一个应答信号,但要产生一个停止信号。

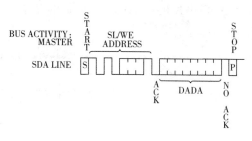

图 24-13 立即地址读时序

(2)随机地址读取

如图24-14所示为24C02随机地址读时序图。随机读操作允许主器件对寄存器的任意字节进行读操作，主器件首先通过发送起始信号、从器件地址和它想读取的字节数据的地址执行一个伪写操作。

在 24C02 应答之后，主器件重新发送起始信号和从器件地址，此时 R/W 位置 1，24C02 响应并发送应答信号，然后输出所要求的一个 8 位字节数据，主器件不发送应答信号但产生一个停止信号。

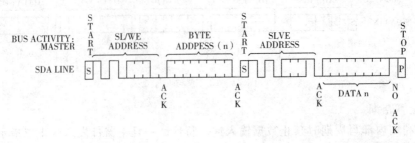

图 24-14　随机地址读时序

（3）顺序地址读取

图 24-15 为 24C02 顺序地址读时序图。顺序读操作可通过立即读或选择性读操作启动。在 24C02 发送完一个 8 位字节数据后，主器件产生一个应答信号来响应，告知 24C02 主器件要求更多的数据，对应每个主机产生的应答信号 24C02 将发送一个 8 位数据字节。当主器件不发送应答信号而发送停止位时结束此操作。

从 24C02 输出的数据按顺序由 N 到 N+1 输出。读操作时地址计数器在 24C02 整个地址内增加，这样整个寄存器区域在可在一个读操作内全部读出。当读取的字节超过 E（24C02，E=255）计数器将翻转到零并继续输出数据字节。

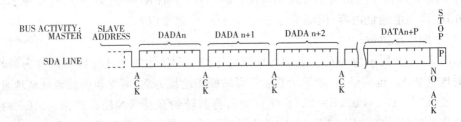

图 24-15　顺序地址读时序

现在以一个实例程序来实现单片机对 24C02 的数据的存储。

二、汇编语言源程序

```
;********************************************************
;项目名称:I²C 总线存储器 AT24C02 读写
;功能:用串口发送指令读出或写入 00-7FH 单元的内容,显示结果在 LCD 上显示
;读出(16 进制数):AA 地址
;写入(16 进制数):BB 地址 内容
;********************************************************
SCK     BIT     P3.3        ;I2C 的时钟线
SDA     BIT     P3.4        ;I2C 的数据线
```

任务24　I²C总线存储器读写

```
;读写I²C总线器件要用到的寄存器
        ERRFLAG     BIT     00H
        TEMP1       EQU     1AH
        DELAYCOUNT  EQU     1BH
        ADDREHI     EQU     1CH         ;读写的地址
        ADDRELO     EQU     1DH
        WRITE_DATA  EQU     1EH         ;单字节写入的数据
        READ_DATA   EQU     1FH         ;单字节读出的数据

        E           BIT     P2.2
        RW          BIT     P2.1
        RS          BIT     P2.0

        LCDPORT     EQU     P0
        CMD_BYTE    EQU     30H
        DAT_BYTE    EQU     31H

        READBIT     BIT     01H         ;读某个单元
        WRITEBIT    BIT     02H         ;写某个单元
        RS232OKBIT  BIT     03H         ;一个完整的串口指令完
        RECSUM      EQU     32H         ;要从232收的字字数

            ORG 0000H
            AJMP MAIN
            ORG 000BH
            AJMP TIME0_1
            ORG 0023H
RS232:      CLR RI
            MOV A,SBUF
            CJNE A,#0AAH,RS232_1
            MOV RECSUM,#01H             ;是读EEP后面只要一个地址就可
            MOV R0,#40H
            SETB READBIT
            AJMP RS232_RE
RS232_1:CJNE A,#0BBH,RS232_2
            MOV RECSUM,#02H
            MOV R0,#40H
            SETB WRITEBIT
            AJMP RS232_RE
RS232_2:MOV @R0,A
            INC R0
            DJNZ RECSUM,RS232_RE
            SETB RS232OKBIT
```

RS232_RE:RETI

```
            ORG     0100H
MAIN:       MOV     R0,#20H
CLR0:       MOV     @R0,#00H
            INC     R0
            CJNE    R0,#70H,CLR0
            MOV     ADDRELO,#00H
            LCALL   READ_BYTE
            LCALL   INITLCD
            MOV     CMD_BYTE,#80H       ;设置 DDRAM 的地址
            LCALL   WRITE_CMD
            LCALL   DELAY0
            MOV     DAT_BYTE,#"A"
            LCALL   WRITE_DAT
            MOV     DAT_BYTE,#"D"
            LCALL   WRITE_DAT
            MOV     DAT_BYTE,#"D"
            LCALL   WRITE_DAT
            MOV     DAT_BYTE,#"R"
            LCALL   WRITE_DAT
            MOV     DAT_BYTE,#"E"
            LCALL   WRITE_DAT
            MOV     DAT_BYTE,#":"
            LCALL   WRITE_DAT
            LCALL   DISPEEPAD

            MOV     CMD_BYTE,#0C0H      ;设置 DDRAM 的地址
            LCALL   WRITE_CMD
            MOV     DAT_BYTE,#"M"
            LCALL   WRITE_DAT
            MOV     DAT_BYTE,#"E"
            LCALL   WRITE_DAT
            MOV     DAT_BYTE,#"M"
            LCALL   WRITE_DAT
            MOV     DAT_BYTE,#"O"
            LCALL   WRITE_DAT
            MOV     DAT_BYTE,#"R"
            LCALL   WRITE_DAT
            MOV     DAT_BYTE,#"Y"
            LCALL   WRITE_DAT
            MOV     DAT_BYTE,#":"
            LCALL   WRITE_DAT
```

```
            LCALL      DISPEEPMO

            MOV        TH0,#0FCH              ;1MS
            MOV        TL0,#67H
            MOV        TH1,#0FDH              ;9600 波特
            MOV        TL1,#0FDH
            MOV        TMOD,#21H
            MOV        SCON,#50H
            SETB       TR0
            SETB       TR1
            MOV        IE,#92H
            SETB       REN
MAIN1:      ORL        PCON,#01H
            AJMP       MAIN1


            ORG        0200H
TIME0_1:    MOV        TH0,#0FCH
            MOV        TL0,#67H
            JBC        RS232OKBIT,TODP
            AJMP       TIME0_RE
TODP:       JBC        READBIT,READ_EEP
            JBC        WRITEBIT,WRITE_EEP
            AJMP       TIME0_RE


TIME0_RE:RETI

READ_EEP:   MOV        ADDRELO,40H
            LCALL      READ_BYTE
            LCALL      DISPEEPAD
            LCALL      DISPEEPMO
            AJMP       TIME0_RE
WRITE_EEP:  MOV        ADDRELO,40H
            MOV        WRITE_DATA,41H
            MOV        READ_DATA,41H
            LCALL      WRITE_BYTE
            LCALL      DISPEEPAD
            LCALL      DISPEEPMO
            AJMP       TIME0_RE

;显示 EEP 地址
DISPEEPAD:  MOV        CMD_BYTE,#87H          ;设置 DDRAM 的地址
            LCALL      WRITE_CMD
            MOV        A,ADDRELO
```

```
                ANL       A,#0F0H
                SWAP      A
                CJNE      A,#0AH,DPEEPAD_1
DPEEPAD_1:      JNC       DPEEPAD_2
                ADD       A,#30H
                MOV       DAT_BYTE,A
                AJMP      DPEEPAD_3
DPEEPAD_2:      ADD       A,#37H
                MOV       DAT_BYTE,A
DPEEPAD_3:      LCALL     WRITE_DAT

                MOV       A,ADDRELO
                ANL       A,#0FH
                CJNE      A,#0AH,DPEEPAD_4
DPEEPAD_4:      JNC       DPEEPAD_5
                ADD       A,#30H
                MOV       DAT_BYTE,A
                SJMP      DPEEPAD_6
DPEEPAD_5:      ADD       A,#37H
                MOV       DAT_BYTE,A
DPEEPAD_6:      LCALL     WRITE_DAT
                RET

DISPEEPMO:      MOV       CMD_BYTE,#0C7H        ;设置DDRAM的地址
                LCALL     WRITE_CMD
                MOV       A,READ_DATA
                ANL       A,#0F0H
                SWAP      A
                CJNE      A,#0AH,DPEEPMO_1
DPEEPMO_1:      JNC       DPEEPMO_2
                ADD       A,#30H
                MOV       DAT_BYTE,A
                AJMP      DPEEPMO_3
DPEEPMO_2:      ADD       A,#37H
                MOV       DAT_BYTE,A
DPEEPMO_3:      LCALL     WRITE_DAT

                MOV       A,READ_DATA
                ANL       A,#0FH
                CJNE      A,#0AH,DPEEPMO_4
DPEEPMO_4:      JNC       DPEEPMO_5
                ADD       A,#30H
                MOV       DAT_BYTE,A
```

```
                SJMP        DPEEPMO_6
DPEEPMO_5:      ADD         A,#37H
                MOV         DAT_BYTE,A
DPEEPMO_6:      LCALL       WRITE_DAT
                RET


;LCD1602 要用到的一些子程序
;写命令(入口参数 CMD_BYTE)
WRITE_CMD:      CLR         RS
                CLR         RW
                MOV         A,CMD_BYTE
                MOV         LCDPORT,A
                SETB        E
                NOP
                NOP
                CLR         E
                LCALL       DELAY0
                RET


;写显示数据(入口参数 DAT_BYTE)
WRITE_DAT:      SETB        RS
                CLR         RW
                MOV         A,DAT_BYTE
                MOV         LCDPORT,A
                SETB        E
                NOP
                NOP
                CLR         E
                LCALL       DELAY0
                RET


;LCD 显示初始化
INITLCD:        MOV CMD_BYTE,#30H
                LCALL       WRITE_CMD
                MOV         CMD_BYTE,#30H
                LCALL       WRITE_CMD
                MOV         CMD_BYTE,#30H
                LCALL       WRITE_CMD
                MOV         CMD_BYTE,#38H          ;设定工作方式
                LCALL       WRITE_CMD
                MOV         CMD_BYTE,#0CH          ;显示状态设置
                LCALL       WRITE_CMD
                MOV         CMD_BYTE,#01H          ;清屏
```

```
            LCALL   WRITE_CMD
            MOV     CMD_BYTE,#06H           ;输入方式设置
            LCALL   WRITE_CMD
            RET

;延时子程序
DELAY0:     MOV     R5,#0A0H
DELAY1:     NOP
            DJNZ    R5,DELAY1
            RET

;延时子程序
DELAY:      MOV     A,DELAYCOUNT
            JZ      DELAY_RE
            DEC     DELAYCOUNT
            MOV     R7,#0C8H
DELAY_1:    NOP
            NOP
            NOP
            DJNZ    R7,DELAY_1
            LJMP    DELAY
DELAY_RE:   RET

;24CXX 要用到的子程序
;单字节写(入口 ADDREHI,ADDRELO,WRITE_DATA)
WRITE_BYTE:LCALL    STARTI2C
            JB      ERRFLAG,WRITE_B_RE      ;不是空闲返回
            MOV     A,#0A0H
            LCALL   SHOUT_DATA
            JB      ERRFLAG,WRITE_B_STOP    ;出错停止
;MOV        A,ADDREHI   ;送高位地址
;LCALL      SHOUT_DATA
;JB         ERRFLAG,WRITE_B_STOP
            MOV     A,ADDRELO               ;送低位地址
            LCALL   SHOUT_DATA
            JB      ERRFLAG,WRITE_B_STOP
            MOV     A,WRITE_DATA
            LCALL   SHOUT_DATA
            JB      ERRFLAG,WRITE_B_STOP
            CLR     ERRFLAG                 ;正常清出错位
WRITE_B_STOP:LCALL  STOPI2C
            MOV     DELAYCOUNT,#0AH
            LCALL   DELAY
```

WRITE_B_RE: RET

;单字节写没有延时
```
WRITE_BYTE0:   LCALL    STARTI2C
               JB       ERRFLAG,WRITE_B_RE0     ;不是空闲返回
               MOV      A,#0A0H
               LCALL    SHOUT_DATA
               JB       ERRFLAG,WRITE_B_STOP0   ;出错停止
               MOV      A,ADDRELO               ;送低位地址
               LCALL    SHOUT_DATA
               JB       ERRFLAG,WRITE_B_STOP0
               MOV      A,WRITE_DATA
               LCALL    SHOUT_DATA
               JB       ERRFLAG,WRITE_B_STOP0
               CLR      ERRFLAG                 ;正常清出错位
WRITE_B_STOP0: LCALL    STOPI2C
WRITE_B_RE0:   RET
```

;单字节读(入口 ADDREHI,ADDRELO;出口 READ_DATA)
```
READ_BYTE:     LCALL    STARTI2C
               JB       ERRFLAG,READ_RE
               MOV      A,#0A0H
               LCALL    SHOUT_DATA
               JB       ERRFLAG,READ_STOP
;MOV           A,ADDREHI
;LCALL         SHOUT_DATA
;JB            ERRFLAG,READ_STOP
               MOV      A,ADDRELO
               LCALL    SHOUT_DATA
               JB       ERRFLAG,READ_STOP
               LCALL    READ_CURRENT
               LJMP     READ_RE
READ_STOP:     LCALL    STOPI2C
READ_RE:       RET

READ_CURRENT:  LCALL    STARTI2C
               JB       ERRFLAG,RE_CURRENTRE
               MOV      A,#0A1H
               LCALL    SHOUT_DATA
               JB       ERRFLAG,RE_CURRENTST
               LCALL    SHIN_DATA
               LCALL    NAK
               CLR      ERRFLAG
```

```
RE_CURRENTST: LCALL      STOPI2C
RE_CURRENTRE: RET

SHOUT_DATA:   MOV        R6,#08H
SHOUT_DATA1:  RLC        A
              MOV        SDA,C
              NOP
              SETB       SCK
              NOP
              NOP
              NOP
              NOP
              CLR        SCK
              DJNZ       R6,SHOUT_DATA1
              SETB       SDA
              NOP
              NOP
              SETB       SCK
              NOP
              NOP
              NOP
              NOP
              MOV        C,SDA
              MOV        ERRFLAG,C
              CLR        SCK
              RET

SHIN_DATA:    SETB       SDA
              MOV        R6,#08H
SHIN_DATA1:   NOP
              NOP
              NOP
              SETB       SCK
              NOP
              NOP
              MOV        C,SDA
              RLC        A
              NOP
              CLR        SCK
              DJNZ       R6,SHIN_DATA1
              MOV        READ_DATA,A
              RET
```

```
STARTI2C:       SETB    SDA
                SETB    SCK
                NOP
                JNB     SDA,STARTI2C_ERR
                JNB     SCK,STARTI2C_ERR
                NOP
                CLR     SDA
                NOP
                NOP
                NOP
                NOP
                CLR     SCK
                CLR     ERRFLAG
                LJMP    STARTI2C_RE
STARTI2C_ERR:   SETB    ERRFLAG
STARTI2C_RE:    RET

STOPI2C:        CLR SDA
                NOP
                NOP
                SETB    SCK
                NOP
                NOP
                NOP
                NOP
                SETB    SDA
                RET

ACK:            CLR     SDA
                NOP
                NOP
                SETB    SCK
                NOP
                NOP
                NOP
                NOP
                CLR     SCK
                RET

NAK:            SETB    SDA
                NOP
                NOP
                SETB    SCK
```

```
                NOP
                NOP
                NOP
                NOP
                CLR         SCK
                RET

                END
```

三、C语言源程序

```c
// *************************************************************************
//项目名称:I²C 总线存储器读写
//功能:用串口发送指令读出或写入 00 - 7FH 单元的内容,显示结果在 LCD 上显示
//读出(16 进制数):AA 地址
//写入(16 进制数):BB 地址 内容
// *************************************************************************
#include <reg51.h>
#include <intrins.h>
#define Uchar unsigned char
#define Uint  unsigned int
#define SomeNOP();  _nop_();_nop_();_nop_();


//定义变量
Uchar recip,recsum,recdata[2];          //接收指针,接收个数,用户标志,接收存放数组
Uchar bdata myflag;
sbit readbit = myflag^0;                //读 EEPROM 标志
sbit writebit = myflag^1;               //写 EEPROM 标志
sbit rs232okbit = myflag^2;             //接收完一拍完整的指令

/* I²C 要用到的口线定义 */
sbit SDA = P3^4;
sbit SCK = P3^3;

/* LCD 控制的口线定义 */
sbit E = P2^2;
sbit RW = P2^1;
sbit RS = P2^0;

//LCD 驱动要用到的一些子程序
//延时 40us
void Delay(unsigned int t)              // delay 40us
{
    for(;t! = 0;t - -);
```

}
//往 LCD 写入命令子程序
```
void SendCommandByte(unsigned char ch)
{
    RS = 0;
    RW = 0;
    P0 = ch;
    E = 1;
    Delay(1);
    E = 0;
    Delay(100);    //delay 40us
}
```

//往 LCD 写入数据子程序
```
void SendDataByte(unsigned char ch)
{
    RS = 1;
    RW = 0;
    P0 = ch;
    E = 1;
    Delay(1);
    E = 0;
    Delay(100); //delay 40us
}
```

//LCD 初始化子程序
```
void InitLcd()
{
    SendCommandByte(0x30);
    SendCommandByte(0x30);
    SendCommandByte(0x30);
    SendCommandByte(0x38);      //设置 LCD 工作方式
    SendCommandByte(0x0c);      //显示 LCD 状态设置
    SendCommandByte(0x01);      //LCD 清屏
    SendCommandByte(0x06);      //LCD 输入方式设置
}
```

/* I²C 要用到的一些函数 */
/* 延时子程序 */
```
void Delay_1ms(Uchar time)
{
    Uchar j;
    while(time - -)
```

```c
    {
        for(j = 163;j! = 0;j - -)
        {;}
    }
}

/* I²C 的启动程序 */
void I2CStart(void)
{
    SDA = 1;
    SomeNOP( );
    SCK = 1;
    SomeNOP( );
    SDA = 0;
    SomeNOP( );
    SCK = 0;
    SomeNOP( );
}

/* I²C 的停止程序 */
void I2CStop(void)
{
    SDA = 0;
    SomeNOP( );
    SCK = 1;
    SomeNOP( );
    SDA = 1;
    SomeNOP( );
}

//Acknowledge 信号
void ACK(void)                              //Acknowledge 信号
{
    SDA = 0;
    SomeNOP( );
    SCK = 1;
    SomeNOP( );
    SCK = 0;
    SomeNOP( );
}

//没有 Acknowledge 信号
void NACK(void)                             //没有 Acknowledge 信号
```

任务 24 I²C 总线存储器读写

```c
{
    SDA = 1;
    SomeNOP( );
    SCK = 1;
    SomeNOP( );
    SCK = 0;
    SomeNOP( );
}

//在一定的时间内自动确定 ACK 的信号//
void AutoACK(void)                              //在一定的时间内自动确定 ACK 的信号
{
    Uchar i = 0;
    SCK = 1;
    SomeNOP();
    while((SDA == 1)&&(i<255))i++;
    SCK = 0;
    SomeNOP();
}

/* I²C 写数据程序(向数据线上写一个 BYTE) */
void Writex(Uchar j)
{
    Uchar i,temp;
    temp = j;
    for(i=0;i<8;i++)//串行移位,先送高位,再送低位//
    {
        temp = temp<<1;
        SCK = 0;
        SomeNOP( );
        SDA = CY;
        SomeNOP( );
        SCK = 1;
        SomeNOP( );
    }
    SCK = 0;
    SomeNOP( );
    SDA = 1;
    SomeNOP( );
}

/* I²C 读数据程序(从数据线上读出一个 BYTE)返回值为读出的数据(BYTE 型) */
Uchar Readx(void)
```

```c
{
    Uchar i,j,k = 0;
    SCK = 0;
    SomeNOP( );
    SDA = 1;
    for(i = 0;i<8;i + +)                        // 串行移位,先读出高位,后读出低位
    {
        SomeNOP( );
        SCK = 1;
        SomeNOP( );
        if(SDA = = 1)j = 1;
        elsej = 0;
        k = (k<<1)|j;
        SCK = 0;
    }
    SomeNOP( );
    return  (k);//返回读出的值//
}

//* * * * * * *以下为对24CXX系列的读写程序* * * * * * * * * * * * * * * * * * *//
//* * * * * * * * * * *24CXX的device address写为0A0H,读为0A1H* * * * * * * *//
/*某个地址写数据*/
void X24_Write(Uchar Addresslow,Uchar Write_date)
{
    I2CStart( );
    Writex(0xA0);
    AutoACK( );
    Writex(Addresslow);
    AutoACK( );
    Writex(Write_date);
    AutoACK( );
    I2CStop( );
    Delay_1ms(10);
}

/*对某个地址读操作,函数值为读出的数据*/
Uchar X24_Read(Uchar Addresslow)
{
    Uchar i;
    I2CStart( );
    Writex(0xA0);
    AutoACK( );
    Writex(Addresslow);
```

```
        AutoACK( );
        I2CStart( );
        Writex(0xA1);
        AutoACK( );
        i = Readx( );
        //ACK( );                              //注意:这里不能有 ACK 应答信号//
        I2CStop( );
        Delay_1ms(0);
        return  (i);
}

//以下为串行口中断程序,用于接收串行数据
void rs232(void)interrupt 4
{
        if(RI = = 1)
        {
           RI = 0;
           if(SBUF = = 0xaa)
           {
              recsum = 0x01;
              readbit = 1;
              recip = 0x00;
           }
           else
           {
              if(SBUF = = 0xbb)
              {
                 recsum = 0x02;
                 writebit = 1;
                 recip = 0x00;
              }
              else
              {
                 recdata[recip] = SBUF;
                 recip + + ;
                 recsum - - ;
                 if(recsum = = 0x00){rs232okbit = 1;}
              }
           }
        }
}

//以下为定时器中断程序
```

```c
void time0(void)interrupt 1
{
    Uchar temp,k;
    TH0 = 0xfc;
    TL0 = 0x67;
    if(rs232okbit = = 1)
    {
        rs232okbit = 0;
        if(readbit = = 1)
        {
            readbit = 0;
            temp = X24_Read(recdata[0]);
            SendCommandByte(0x87);
            k = (recdata[0]>>4)&0x0f;
            if(k<0x0a){k = k + 0x30;}
            else{k = k + 0x37;}
            SendDataByte(k);
            k = recdata[0]&0x0f;
            if(k<0x0a){k = k + 0x30;}
            else{k = k + 0x37;}
            SendDataByte(k);
            SendCommandByte(0xC7);
            k = (temp>>4)&0x0f;
            if(k<0x0a){k = k + 0x30;}
            else{k = k + 0x37;}
            SendDataByte(k);
            k = temp&0x0f;
            if(k<0x0a){k = k + 0x30;}
            else{k = k + 0x37;}
            SendDataByte(k);
        }
        if(writebit = = 1)
        {
            writebit = 0;
            SendCommandByte(0x87);
            k = (recdata[0]>>4)&0x0f;
            if(k<0x0a){k = k + 0x30;}
            else{k = k + 0x37;}
            SendDataByte(k);
            k = recdata[0]&0x0f;
            if(k<0x0a){k = k + 0x30;}
            else{k = k + 0x37;}
            SendDataByte(k);
```

```
            SendCommandByte(0xC7);
            k=(recdata[1]>>4)&0x0f;
            if(k<0x0a){k=k+0x30;}
            else{k=k+0x37;}
            SendDataByte(k);
            k=recdata[1]&0x0f;
            if(k<0x0a){k=k+0x30;}
            else{k=k+0x37;}
            SendDataByte(k);
            X24_Write(recdata[0],recdata[1]);
        }
    }
}

//主程序
void main(void)
{
    InitLcd();
    SendCommandByte(0x80);
    Delay(2);
    SendDataByte('A');
    SendDataByte('D');
    SendDataByte('D');
    SendDataByte('R');
    SendDataByte('E');
    SendDataByte(':');
    SendCommandByte(0xC0);
    Delay(2);
    SendDataByte('M');
    SendDataByte('E');
    SendDataByte('M');
    SendDataByte('O');
    SendDataByte('R');
    SendDataByte('Y');
    SendDataByte(':');
    TH0=0xfc;
    TL0=0x67;
    TH1=0xfd;
    TL1=0xfd;
    TMOD=0x21;
    SCON=0x50;
    TR0=1;TR1=1;
    IE=0x92;
```

```
REN = 1;
readbit = 1;
rs232okbit = 1;
while(1)              //等待中断
{ }
}
```

实训考核

本课程改革传统的闭卷或开卷考核,而采用过程考核为主的多元化考核方式,考核分为理论考核、职业道德考核和技能考核三部分,各部分所占比例见表24-1、表24-2、表24-3。

表24-1 理论考核和职业素质考核形式及所占比例

序号	名称		比例	得分
一	理论考核	过程作业文件	个人自评	10%	
			组内互评	20%	
			小组互评	30%	
			老师评定	20%	
		课堂提问、解答		10%	
		项目汇报		10%	
		小计		100%	
二	职业素质	职业道德,工作作风		40%	
		小组沟通协作能力		40%	
		创新能力		20%	
		小计		100%	

表24-2 技能考核内容及比例

姓名		班级		小组		总得分	
序号	考核项目	考核内容及要求		配分	评分标准	考核环节	得分
1	①安全文明生产 ②安全操作规范	着装规范		20%	现场考评	实施	
		安全用电					
		布线规范、合理					
		工具摆放整齐					
		工具及仪器仪表使用规范、摆放整齐					
		任务完成后,进行场地整理,保持场地清洁、有序					

(续表)

序号	考核项目	考核内容及要求	配分	评分标准	考核环节	得分
2	实训态度	不迟到、早退、旷课	10%	现场考评	六步	
		实训过程认真负责				
		组内主动沟通、协作,小组间互助				
3	系统方案制定	工作流程正确合理	10%	现场考评	计划决策	
		方案合理				
		选用指令是否合理				
		电路图正确				
	编程能力	独立完成程序	10%	现场考评	决策	
		程序简单、可靠				
4	操作能力	正确输入程序并进行程序调试	20%	现场考评	实施	
		根据电路图正确接线				
		根据系统功能进行正确操作演示				
5	工艺	接线美观	10%	现场考评	实施	
		线路工作可靠				
6	实践效果	系统工作可靠	10%	现场考评	检查	
		满足工作要求				
		创新				
		按规定的时间完成项目				
7	汇报总结	工作总结,PPT汇报	5%	现场考评	评估	
		填写自我检查表及反馈表				
8	技术文件制作整理	技术文件制作整理能力	5%	现场考评	评估	
		合计	100%			

表24-3 各部分考核占课程考核的比例

考核项目	理论考核	技能考核	职业素质考核	合 计
比 例	30%	50%	20%	100%
分 值	30	50	20	100
实际得分				

任务 25　DS18B20 温度控制

实训任务

通过读取数字温度传感器的值,对比设定值,当温度值大于设定值时,继电器输出。温度设定值可以通过按键 K1、K2 来实现加减,并且在 LCD1602 上显示出来。同时测量的温度值也显示在 LCD1602 上。

实训设备

1. 设备
PC 机(安装 wave 编程软件、Keil C51 软件)、单片机实验板。
2. 工具及材料
工作对象:电工电子工具、电子元器件和辅助材料、仿真器、编程器。
工作工具:单片机控制电路原理图、实训指导书、项目任务单、工作记录单、项目检查单、各种电工仪表、常用电工工具和拆装工具、量具、相关电子手册。

硬件设计

主控模块采用 ATMEL 公司生产的 AT89S52 单片机,温度传感器采用 DALLAS 公司生产的 DS18B20,温度传感器模块与单片机的接口电路原理图如图 25-1 所示。

软件设计

一、DS18B20 基本知识

DS18B20 数字温度计是 DALLAS 公司生产的 1-Wire,即单总线器件,具有线路简单,体积小的特点。因此用它来组成一个测温系统,具有线路简单,在一根通信线,可以挂很多这样的数字温度计,十分方便。
1. DS18B20 产品的特点
(1)只要求一个端口即可实现通信。
(2)在 DS18B20 中的每个器件上都有独一无二的序列号。

任务 25 DS18B20 温度控制

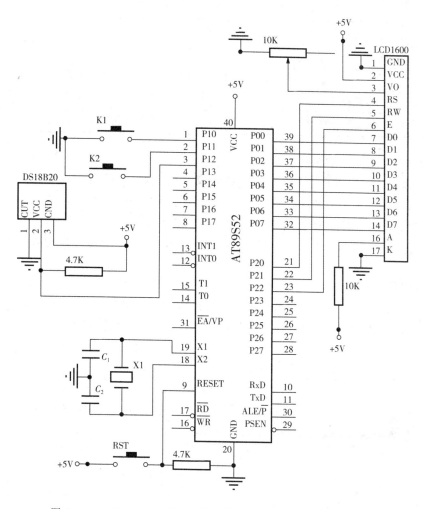

图 25-1 DS18B20 温度传感器模块与单片机的接口电路原理图

(3)实际应用中不需要外部任何元器件即可实现测温。
(4)测量温度范围在 -55℃ 到 +125℃ 之间。
(5)数字温度计的分辨率用户可以从 9 位到 12 位选择。
(6)内部有温度上、下限告警设置。

图 25-2 DS18B20 的底视图

2. DS18B20 的引脚介绍

TO-92 封装的 DS18B20 的引脚排列见图 25-2,其引脚功能描述见表 25-1。

表 25-1　DS18B20 详细引脚功能描述

序号	名称	引脚功能描述
1	GND	地信号
2	DQ	数据输入/输出引脚。开漏单总线接口引脚。当被用着在寄生电源下，也可以向器件提供电源
3	VDD	可选择的 VDD 引脚。当工作于寄生电源时，此引脚必须接地

3．DS18B20 的使用方法

由于 DS18B20 采用的是 1－Wire 总线协议方式，即在一根数据线实现数据的双向传输，而对 AT89S51 单片机来说，硬件上并不支持单总线协议，因此，我们必须采用软件的方法来模拟单总线协议时序来完成对 DS18B20 芯片的访问。

由于 DS18B20 是在一根 I/O 线上读写数据，因此，对读写的数据位有着严格的时序要求。DS18B20 有严格的通信协议来保证各位数据传输的正确性和完整性。该协议定义了几种信号的时序：初始化时序、读时序、写时序。所有时序都是将主机作为主设备，单总线器件作为从设备。而每一次命令和数据的传输都是从主机主动启动写时序开始，如果要求单总线器件回送数据，在进行写命令后，主机需启动读时序完成数据接收。数据和命令的传输都是低位在先。

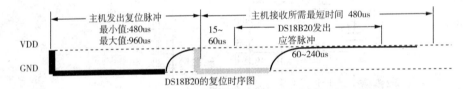

图 25-3　DS18B20 的复位时序图

(1)DS18B20 的读时序

对于 DS18B20 的读时序分为读 0 时序和读 1 时序两个过程。

对于 DS18B20 的读时隙是从主机把单总线拉低之后，在 15 秒之内就得释放单总线，以让 DS18B20 把数据传输到单总线上。DS18B20 在完成一个读时序过程，至少需要 60us 才能完成。

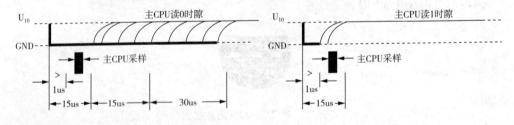

图 25-4　DS18B20 的读时序图

(2)DS18B20 的写时序

对于 DS18B20 的写时序仍然分为写 0 时序和写 1 时序两个过程。

对于 DS18B20 写 0 时序和写 1 时序的要求不同。

当要写 0 时序时,单总线要被拉低至少 60us,保证 DS18B20 能够在 15us 到 45us 之间能够正确地采样 I/O 总线上的"0"电平。

当要写 1 时序时,单总线被拉低之后,在 15us 之内就得释放单总线。

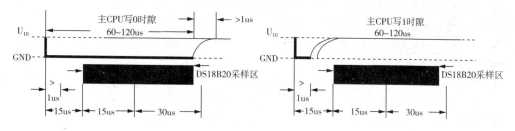

图 25-5　DS18B20 的写时序图

(3)读写 DS18B20 的指令说明

见表 25-2。

表 25-2　DS18B20 的指令说明

Command	Description	Protocol	1—Wire Bus Activity After Command is Issued	Notes
TEMPERATURE CONVERSION COMMANDS				
Convert T	Initiates temperature conversion.	44h	DS18B20 transmits conversion status to master (not applicable for parasite-powered DS18B20s)	1
MEMORY COMMANDS				
Read Scratchpad	Reads the entire scratchpad including the CRC byte.	BEh	DS18B20 transmits up to 9 data bytes to master	2
Write Scratchpad	Writes data into scratchpad bytes 2,3, and 4 (T_H, T_L, and configuration registers).	4Eh	Master transmits 3 data bytes to DS18B20	3
Copy Scratchpad	Copies T_H, T_L, and configuration register data from the scratchpad to EEPROM.	48h	None	1
Recall E^2	Recalls T_H, T_L, and configuration register data from EEPROM to the scratchpad.	B8h	DS18B20 transmits recall status to master	
Read Power Supply	Signals DS18B20 power supply mode to the master.	B4h	DS18B20 transmits supply status to master	

二、C语言源程序

```c
// ******************************************************************************
//项目名称:DS18B20 温度控制
//功能:通过读取数字温度传感器的值,对比设定值,当温度值大于设定值时,继电器输出。
//温度设定值可以通过按键 K1、K2 来实现加减,并且在 LCD1602 上显示出来;
//同时测量的温度值也显示在 LCD1602 上。
// ******************************************************************************
#include <intrins.h>
#include <reg52.h>
#define Uchar unsigned char
#define Uint  unsigned int
#define SomeNOP( ); _nop_( );
sbit DQ = P3^7;                     //DS18B20 温度传感器
sbit JDQ = P3^6;                    //继电器输出
//LCD1602 口线接法
sbit E = P2^2;
sbit RW = P2^1;
sbit RS = P2^0;

//I2C 总线要用到的口线定义
sbit SDA = P3^4;                    //I2C 总线数据信号
sbit SCK = P3^3;                    //I2C 总线时钟信号


//以下为函数声明
void TempDelay(unsigned char us);   //延时函数
void Ds18b20_rst(void);             //DS18B20 复位
void WriteByte(unsigned char wrdata);//往 DS18B20 写入数据
unsigned char ReadByte(void);       //从 DS18B20 读出数据
void Delay(unsigned int t);         //delay 40us
void SendCommandByte(unsigned char c);//往 LCD1602 写入指令
void SendDataByte(unsigned char d); //往 LCD1602 写入数据
void InitLcd( );                    //LCD1602 初始化

//I2C 驱动要用到的一些函数
void Delay_1ms(Uchar time);
void I2CStart(void);
void I2CStop(void);
void AutoACK(void);                 //在一定的时间内自动确定 ACK 的信号
void Writex(Uchar j);
Uchar Readx(void);
void X24_Write(Uchar Addresslow,Uchar Write_date);
Uchar X24_Read(Uchar Addresslow);
```

```c
void ramini(void);
void time0(void);

union
{
    unsigned int iInt;
    unsigned char cChar[2];
}TMP;

unsigned int nowwd,highwd;           //当前的温度,设定报警
unsigned char keyscan,teml,temh;
unsigned char dssttime,svtime;
unsigned char bdata myflag1;
sbit keybit   = myflag1^0;           //按键标志
sbit dsokbit  = myflag1^1;           //DS18B20 在线
sbit dsstbit  = myflag1^2;
sbit dsfubit  = myflag1^3;
sbit newwdbit = myflag1^4;
sbit newjrbit = myflag1^5;
sbit setjrbit = myflag1^6;

void main(void)
{
    TH0 = 0xf8;
    TL0 = 0xce;
    TMOD = 0x01;
    TR0 = 1;
    keyscan = 50;
    InitLcd();
    ramini();
    SendCommandByte(0x80);           //设置 DDRAM 地址
    SendDataByte('T');
    SendDataByte('e');
    SendDataByte('m');
    SendDataByte('p');
    SendDataByte(':');

    SendCommandByte(0xC0);           //设置 DDRAM 地址
    SendDataByte('S');
    SendDataByte('e');
    SendDataByte('t');
    SendDataByte(' ');
    SendDataByte(':');
```

```c
    IE = 0x82;
    newjrbit = 1;
    while(1){ }
}

void time0(void)interrupt 1
{
    unsigned char k,p;
    TH0 = 0xf8;                    //2ms
    TL0 = 0xce;
    if(keyscan - - = = 0)
    {
        keyscan = 50;
        k = P1;k = k&0x03;
        if(k! = 0x03)
        {
           switch(k)
           {
              case 0x02:
              if(keybit = = 0){keybit = 1;highwd + + ;setjrbit = 1;svtime = 30;newjrbit = 1;}break;
              case 0x01:
              if(keybit = = 0){keybit = 1;highwd - - ;setjrbit = 1;svtime = 30;newjrbit = 1;}break;
           }
        }
        else{keybit = 0;}
        if(setjrbit = = 1)
        {
            if(svtime - - = = 0)
            {
                TMP.iInt = highwd;
                X24_Write(0x10,TMP.cChar[0]);
                X24_Write(0x11,TMP.cChar[1]);
                setjrbit = 0;
            }
        }
        if(dsstbit = = 0)          //采集温度值
        {
            Ds18b20_rst();
            if(dsokbit = = 1)
            {
                WriteByte(0xCC);//忽略 ROM 匹配
                WriteByte(0x44);//发送温度转化命令
                dsstbit = 1;dssttime = 0x07;
```

```
        }
    }
    else
    {
        if(dssttime- - = = 0)
        {
            Ds18b20_rst();
            if(dsokbit = = 1)
            {
                WriteByte(0xCC);
                WriteByte(0xBE);
                teml = ReadByte();
                temh = ReadByte();
                k = temh&0xfc;
                if(k! = 0xfc)
                {
                    k = (temh<<4)&0xf0;p = (teml>>4)&0x0f;
                    k = k|p;

                    nowwd = k * 10;
                    k = teml&0x0f;k = k * 10;p = k/16;
                    if((k%16)>5){p = p + 1;}
                    nowwd = nowwd + p;
                    dsstbit = 0;dsfubit = 0;
                }
                else
                {
                    k = (temh<<4)&0xf0;p = (teml>>4)&0x0f;
                    k = k|p;
                    k = (~k);
                    nowwd = k * 10;
                    k = (teml&0x0f)|0xf0;k = ~k;k = k * 10;p = k/16;
                    if((k%16)>5){p = p + 1;}
                    nowwd = nowwd + p;
                    dsstbit = 0;dsfubit = 1;
                }
                newwdbit = 1;
            }
        }
    }
}

if(newwdbit = = 1)
```

```c
    {
        newwdbit = 0;
        if(nowwd<highwd)    { JDQ = 1; }
        else                { JDQ = 0; }
        SendCommandByte(0x87);    //设置 DDRAM 地址

        k = (Uchar)(nowwd/1000);
        k = k + 0x30;
        SendDataByte(k);
        k = (Uchar)((nowwd%1000)/100);
        k = k + 0x30;
        SendDataByte(k);
        k = (Uchar)((nowwd%100)/10);
        k = k + 0x30;
        SendDataByte(k);
        SendDataByte('.');
        k = (Uchar)(nowwd%10);
        k = k + 0x30;
        SendDataByte(k);
    }

    if(newjrbit = = 1)
    {
        newjrbit = 0;
        SendCommandByte(0xc7);    //设置 DDRAM 地址

        k = (Uchar)(highwd/1000);
        k = k + 0x30;
        SendDataByte(k);
        k = (Uchar)((highwd%1000)/100);
        k = k + 0x30;
        SendDataByte(k);
        k = (Uchar)((highwd%100)/10);
        k = k + 0x30;
        SendDataByte(k);
        SendDataByte('.');
        k = (Uchar)(highwd%10);
        k = k + 0x30;
        SendDataByte(k);
    }
}

//DS18B20 要用到的一些函数
```

任务 25　DS18B20 温度控制

```c
void TempDelay(unsigned char us)//延时
{
    while(us--);
}

void Ds18b20_rst(void)           //DS18B20 复位
{
    DQ = 1;
    _nop_();
    DQ = 0;
    TempDelay(86);
    _nop_();
    DQ = 1;
    TempDelay(14);
    _nop_();
    _nop_();
    _nop_();
    if(DQ==0){dsokbit=1;}
    else{dsokbit=0;}
    TempDelay(20);
    _nop_();
    _nop_();
    DQ = 1;
}

//向 DS18B20 写入一字节
void WriteByte(unsigned char wrdat)
{
    unsigned char i;
    for(i=0;i<8;i++)
    {
        DQ = 0;
        _nop_();
        DQ = wrdat&0x01;
        TempDelay(5);
        _nop_();
        _nop_();
        DQ = 1;
        wrdat>>=1;
    }
}

unsigned char ReadByte(void)      //读 DS18B20 的一个字节
```

```c
{
    unsigned char i,u=0;
    for(i=0;i<8;i++)
    {
        DQ=0;
        u>>=1;
        DQ=1;
        if(DQ==1)
        u|=0x80;
        TempDelay(4);
        _nop_();
    }
    return(u);
}
```

//延时函数
```c
void Delay(unsigned int t)          // delay 40us
{
    for(;t!=0;t--);
}
```

//往LCD1602写入指令函数
```c
void SendCommandByte(unsigned char ch)
{
    RS=0;
    RW=0;
    P0=ch;
    E=1;
    Delay(1);
    E=0;
    Delay(100);                     //delay 40us
}
```

//往LCD1602写入数据函数
```c
void SendDataByte(unsigned char ch)
{   RS=1;
    RW=0;
    P0=ch;
    E=1;
    Delay(1);
    E=0;
    Delay(100);                     //delay 40us
}
```

```c
//LCD1602 初始化函数
void InitLcd()
{
    SendCommandByte(0x30);
    SendCommandByte(0x30);
    SendCommandByte(0x30);
    SendCommandByte(0x38);    //设置工作方式
    SendCommandByte(0x0c);    //显示状态设置
    SendCommandByte(0x01);    //清屏
    SendCommandByte(0x06);    //输入方式设置
}

//I²C 要用到的一些子程序
//延时函数
void Delay_1ms(Uchar time)
{
    Uchar j;
    while(time--)
    {
        for(j=163;j!=0;j--)
        {;}
    }
}

//I²C 起动函数
void I2CStart(void)
{
    SDA = 1;
    SomeNOP();
    SCK = 1;
    SomeNOP();
    SDA = 0;
    SomeNOP();
    SCK = 0;
    SomeNOP();
}

//I²C 停止函数
void I2CStop(void)
{
    SDA = 0;
    SomeNOP();
```

```c
    SCK = 1;
    SomeNOP();
    SDA = 1;
    SomeNOP();
}

//I²C 应答函数
void I2CACK(void)                //Acknowledge 信号
{
    SDA = 0;
    SomeNOP();
    SCK = 1;
    SomeNOP();
    SCK = 0;
    SomeNOP();
}

//I²C 非应答函数
void I2CNACK(void)               //没有 Acknowledge 信号
{
    SDA = 1;
    SomeNOP();
    SCK = 1;
    SomeNOP();
    SCK = 0;
    SomeNOP();
}

//I²C 自动应答函数
void AutoACK(void)               //在一定的时间内自动确定 ACK 的信号
{
    Uchar i = 0;
    SCK = 1;
    SomeNOP();
    while((SDA == 1)&&(i<255))i++;
    SCK = 0;
    SomeNOP();
}

//I²C 写数据程序(向数据线上写一个 BYTE)
void Writex(Uchar j)
{
    Uchar i,temp;
```

```
            temp = j;
            for(i=0;i<8;i++)         //串行移位,先送高位,再送低位
            {
                temp = temp<<1;
                SCK = 0;
                SomeNOP();
                SDA = CY;
                SomeNOP();
                SCK = 1;
                SomeNOP();
            }
            SCK = 0;
            SomeNOP();
            SDA = 1;
            SomeNOP();
}

//I²C 读数据函数(从数据线上读出一个 BYTE)返回值为读出的数据(BYTE 型)
Uchar Readx(void)
{
            Uchar i,j,k=0;
            SCK = 0;
            SomeNOP( );
            SDA = 1;
            for(i=0;i<8;i++)         //串行移位,先读出高位,后读出低位
            {
                SomeNOP( );
                SCK = 1;
                SomeNOP( );
                if(SDA==1)j=1;
                else j=0;
                k = (k<<1)|j;
                SCK = 0;
            }
            SomeNOP( );
            return (k);              //返回读出的值
}

//***************以下为对 24CXX 系列的读写程序****************************//
//*********24CXX 的 device address 为写 0A0H,读为 0A1H*******************//
//某个地址写数据
void X24_Write(Uchar Addresslow,Uchar Write_date)
{
```

```
    I2CStart( );
    Writex(0xA0);
    AutoACK( );
    Writex(Addresslow);
    AutoACK( );
    Writex(Write_date);
    AutoACK( );
    I2CStop( );
    Delay_1ms(10);
}

//对某个地址读操作,函数值为读出的数据
Uchar X24_Read(Uchar Addresslow)
{
    Uchar i;
    I2CStart( );
    Writex(0xA0);
    AutoACK( );
    Writex(Addresslow);
    AutoACK( );
    I2CStart( );
    Writex(0xA1);
    AutoACK( );
    i = Readx( );
    //ACK( );                    //注意:这里不能有 ACK 应答信号//
    I2CStop( );
    Delay_1ms(0);
    return (i);
}

//对内存中的数据块写入,入口为写 EEPROM 的地址,内存缓冲中的数据的起始地址(是 IDATA 的数据)
void X24_WritePage(Uchar Addresshigh,Uchar Addresslow,Uchar idata * Writeip)
{
    Uchar i;
    I2CStart( );
    Writex(0xA0);
    AutoACK( );
    Writex(Addresshigh);
    AutoACK( );
    Writex(Addresslow);
    AutoACK( );
    for(i=0;i<16;i++)
    {
```

```
            Writex(*Writeip++);
            AutoACK( );
        }
        I2CStop( );
}

//把EEPROM中的数据读出来存放在内存缓冲中,起始地址为(是IDATA的数据)
void X24_ReadPage(Uchar Addresshigh,Uchar Addresslow,Uchar idata *Readip)
{
        Uchar i;
        I2CStart( );
        Writex(0xA0);
        AutoACK( );
        Writex(Addresshigh);
        AutoACK( );
        Writex(Addresslow);
        AutoACK( );
        I2CStart( );
        Writex(0xA1);
        AutoACK( );
        for(i=0;i<16;i++)           //PAGE读的时候最后一个BYTE不要ACK信号
        {
            *Readip = Readx( );
              Readip = Readip++;
              I2CACK( );
        }
        *Readip = Readx( );
        I2CNACK( );
        I2CStop( );
        Delay_1ms(0);
    }

void ramini(void)
{
    TMP.cChar[0]=X24_Read(0x10);
    TMP.cChar[1]=X24_Read(0x11);
    highwd=TMP.iInt;
}
```

实训考核

本课程改革传统的闭卷或开卷考核,而采用过程考核为主的多元化考核方式,考核分为理论考核、职业道德考核和技能考核三部分,各部分所占比例见表 25-3、表 25-4、表 25-5

所示。

表 25-3 理论考核和职业素质考核形式及所占比例

序号	名 称		比 例	得 分
一	理论考核	个人自评	10%	
		过程作业文件 组内互评	20%	
		小组互评	30%	
		老师评定	20%	
		课堂提问、解答	10%	
		项目汇报	10%	
		小计	100%	
二	职业素质	职业道德,工作作风	40%	
		小组沟通协作能力	40%	
		创新能力	20%	
		小计	100%	

表 25-4 技能考核内容及比例

姓名		班 级		小组		总得分	
序号	考核项目	考核内容及要求		配分	评分标准	考核环节	得分
1	①安全文明生产 ②安全操作规范	着装规范		20%	现场考评	实施	
		安全用电					
		布线规范、合理					
		工具摆放整齐					
		工具及仪器仪表使用规范、摆放整齐					
		任务完成后,进行场地整理,保持场地清洁、有序					
2	实训态度	不迟到、早退、旷课		10%	现场考评	六步	
		实训过程认真负责					
		组内主动沟通、协作,小组间互助					
3	系统方案制定	工作流程正确合理		10%	现场考评	计划决策	
		方案合理					
		选用指令是否合理					
		电路图正确					
	编程能力	独立完成程序		10%	现场考评	决策	
		程序简单、可靠					

(续表)

序号	考核项目	考核内容及要求	配分	评分标准	考核环节	得分
4	操作能力	正确输入程序并进行程序调试	20%	现场考评	实施	
		根据电路图正确接线				
		根据系统功能进行正确操作演示				
5	工艺	接线美观	10%	现场考评	实施	
		线路工作可靠				
6	实践效果	系统工作可靠	10%	现场考评	检查	
		满足工作要求				
		创新				
		按规定的时间完成项目				
7	汇报总结	工作总结,PPT汇报	5%	现场考评	评估	
		填写自我检查表及反馈表				
8	技术文件制作整理	技术文件制作整理能力	5%	现场考评	评估	
	合计		100%			

表 25-5 各部分考核占课程考核的比例

考核项目	理论考核	技能考核	职业素质考核	合计
比例	30%	50%	20%	100%
分值	30	50	20	100
实际得分				

附录 A 特殊功能寄存器

特殊功能寄存器(SFR)也称为专用寄存器，主要用于管理片内和片外的功能部件。MCS—51 系列单片机主要有 21 个特殊功能寄存器，它们离散地分布在内部 RAM 的 80H～FFH 地址中，这些寄存器的功能已做了专门的规定，用户不能修改其结构。附 A 表 1 所示为特殊功能寄存器地址表。

附 A 表 1 特殊功能寄存器地址表

标识符号	寄存器名称	字节地址	是否可以位寻址
B	B 寄存器	F0H	√
ACC	A 累加器	E0H	√
PSW	程序状态字	D0H	√
IP	中断优先级控制寄存器	B8H	√
P3	P3 口寄存器	B0H	√
IE	中断允许控制寄存器	A8H	√
P2	P2 口寄存器	A0H	√
SBUF	串行数据缓冲寄存器	99H	
SCON	串行口控制寄存器	98H	√
P1	P1 口寄存器	90H	√
TH1	定时器 1 高 8 位	8DH	
TH0	定时器 0 高 8 位	8CH	
TL1	定时器 1 低 8 位	8BH	
TL0	定时器 0 低 8 位	8AH	
TMOD	定时器工作方式寄存器	89H	
TCON	定时器控制寄存器	88H	√
PCON	电源控制及波特率选择寄存器	87H	
DPTR	数据指针(16 位)含 DPH 和 DPL	83H,82H	
SP	堆栈指针	81H	
P0	P0 口寄存器	80H	√

附录B MCS-51单片机指令系统

MCS-51单片机指令系统有42种助记符,代表了33种操作功能,这是因为有的功能可以有几种助记符(例如数据传送的助记符有MOV,MOVC,MOVX)。指令功能助记符与操作数各种可能的寻址方式相结合,共构成111种指令。这111种指令中,如果按字节分类,单字节指令49条,双字节指令45条,三字节指令17条。若从指令执行的时间看,单机器周期(12个振荡器周期)指令64条,双机器周期指令45条,四个机器周期指令2条(乘、除)。由此可见,MCS-51单片机指令系统具有存储空间效率高和执行速度快的特点。

按指令的功能,MCS-51单片机指令系统可分为下列5类:

① 数据传送类指令29条(见附B表1)。
② 算术运算类指令24条(见附B表2)。
③ 逻辑运算类指令24条(见附B表3)。
④ 控制转移类指令17条(见附B表4)。
⑤ 位操作类指令17条(见附B表5)。

附B表1 数据传送类指令

指令助记符	机器码(H)	说明	字节数	机器周期数
MOV A,#data	74,data	立即数送累加器;A←#data	2	1
MOV A,direct	E5,direct	直接寻址字节内容送累加器;A←(direct)	2	1
MOV A,Rn	E8~EF	寄存器内容送累加器 ;A←(Rn)	1	1
MOV A,@Ri	E6~E7	间接RAM送累加器 ;A←((Ri))	1	1
MOV Rn,A	F8~FF	累加器送寄存器 ;Rn←A	1	1
MOV Rn,#data	78~7F,data	立即数送寄存器 ;Rn←#data	2	1
MOV Rn,direct	A8~AF,direct	直接寻址字节送寄存器 ;Rn←(direct)	2	2
MOV direct,A	F5,direct	累加器送直接寻址字节;direct←A	2	1
MOV direct,#data	75,direct,data	立即数送直接寻址字节 ;direct←#data	3	2
MOV direct1,direct2	85,direct2,direct1	直接寻址字节送直接寻址字节 ;direct1←(direct2)	3	2
MOV direct,Rn	88~8F,direct	寄存器送直接寻址字节;direct←(Rn)	2	2
MOV direct,@Ri	86~87,direct	间接RAM送直接寻址字节 ;direct←((Ri))	2	2
MOV@Ri,A	F6~F7	累加器送片内RAM ;(Ri)←A	1	1

(续表)

指令助记符	机器码(H)	说 明	字节数	机器周期数
MOV @Ri,#data	76~77,data	立即数送片内 RAM;(Ri)←#data	2	1
MOV @Ri,direct	A6~A7,direct	直接寻址字节送片内 RAM;(Ri)←(direct)	2	2
MOV DPTR,#data16	90,data16	16 位立即数送数据指针;DPRT←#data16	3	2
MOVX A,@Ri	E2~E3	片外 RAM 送累加器(8 位地址);A←((Ri))	1	2
MOVX A,@DPTR	E0	片外 RAM(16 位地址)送累加器;A←((DPTR))	1	2
MOVX @Ri,A	F2~F3	累加器送片外 RAM(8 位地址);((Ri))←A	1	2
MOVX @DPTR,A	F0	累加器送片外 RAM(16 位地址);((DPTR))←A	1	2
MOVC A,@A+DPTR	83	变址寻址字节送累加器(相对 DPTR);A←((A)+(DPTR))	1	2
MOVC A,@A+PC	93	变址寻址字节送累加器(相对 PC);A←((A)+(PC))	1	2
XCH A,direct	C5,direct	直接寻址字节与累加器交换;(A)←→(direct)	2	1
XCH A,Rn	C8~CF	寄存器与累加器交换;(A)←→(Rn)	1	1
XCH A,@Ri	C6~C7	片内 RAM 与累加器交换;(A)←→((Ri))	1	1
XCHD A,@Ri	D6~D7	片内 RAM 与累加器低 4 位交换;(A)$_{3-0}$←→((Ri))$_{3-0}$	1	1
SWAP A	C4	A 半字节交换	1	1
PUSH direct	C0,direct	直接寻址字节压入栈顶;SP←(SP)+1,(SP)←(direct)	2	2
POP direct	D0,direct	栈顶弹至直接寻址字节;direct←((SP)),SP←(SP)-1	2	2

附 B 表 2　算术运算类指令

指令助记符	机器码	说 明	字节数	机器周期数
ADD A,#data	24,#data	立即数送累加器;A←(A)+data	2	1
ADD A,direct	25,direct	直接寻址送累加器;A←(A)+(direct)	2	1
ADD A,Rn	28~2F	寄存器内容送累加器;A←(A)+(Rn)	1	1

(续表)

指令助记符	机器码	说 明	字节数	机器周期数
ADD A,@Ri	26～27	间接寻址 RAM 送累加器；A←(A)+((Ri))	1	1
ADDC A,♯data	34,♯data	立即数加到累加器(带进位)；A←(A)+data+CY	2	1
ADDC A,direct	35,direct	直接寻址加到累加器(带进位)；A←(A)+(direct)+CY	2	1
ADDC A,Rn	38～3F	寄存器加到累加器(带进位)；A←(A)+(Rn)+CY	1	1
ADDC A,@Ri	36～37	间接寻址 RAM 加到累加器(带进位)；A←(A)+((Ri))+CY	1	1
INC A	04	累加器加1；A←(A)+1	1	1
INC direct	05,direct	直接寻址加1；direct←(direct)+1	2	1
INC Rn	08～0F	寄存器加1；Rn←(Rn)+1	1	1
INC @Ri	06～07	间接寻址 RAM 加1；(Ri)←((Ri))+1	1	1
INC DPTR	A3	地址寄存器加1；DPTR←DPTR+1	1	2
DA A	D4	对 A 进行十进制调整	1	1
SUBB A,♯data	94,data	累加器内容减去立即数(带借位)；A←(A)−data−CY	2	1
SUBB A,direct	95,direct	累加器内容减去直接寻址(带借位)；A←(A)−(direct)−CY	2	1
SUBB A,Rn	98～9F	累加器内容减去寄存器内容(带借位)；A←(A)−(Rn)−CY	1	1
SUBB A,@Ri	96～97	累加器内容减去间接寻址(带借位)；A←(A)−((Ri))−CY	1	1
DEC A	14	累加器减1；A←(A)−1	1	1
DEC direct	15,direct	直接寻址地址字节减1；direct←(direct)−1	2	1
DEC Rn	18～1F	寄存器减1；Rn←(Rn)−1	1	1
DEC @Ri	16～17	间接寻址 RAM 减1；(Ri)←((Ri))−1	1	1
MUL AB	A4	累加器 A 和寄存器 B 相乘；AB←(A)*(B)	1	4
DIV AB	84	累加器 A 除以寄存器 B；AB←(A)/(B)	1	4

附 B 表 3 逻辑运算类指令

指令助记符	机器码	说　明	字节数	机器周期数
ANL　A,♯data	54,data	立即数"与"到累加器;A←(A)∧data	2	1
ANL　A,direct	55,direct	直接寻址"与"到累加器;A←(A)∧(direct)	2	1
ANL　A,Rn	58～5F	寄存器"与"到累加器;A←(A)∧(Rn)	1	1
ANL　A,@Ri	56～57	间接寻址 RAM"与"到累加器;A←(A)∧((Ri))	1	1
ANL　direct,A	52,direct	累加器"与"到直接寻址;direct←(direct)∧(A)	2	1
ANL　direct,♯data	53,direct,data	立即数"与"到直接寻址;direct←(direct)∧data	3	2
ORL　A,♯data	44,data	立即数"或"累加器;A←(A)∨data	2	1
ORL　A,direct	45,direct	直接寻址"或"到累加器;A←(A)∨(direct)	2	1
ORL　A,Rn	48～4F	寄存器"或"到累加器;A←(A)∨(Rn)	1	1
ORL　A,@Ri	46～47	间接寻址 RAM"或"到累加器;A←(A)∨((Ri))	1	1
ORL　direct,A	42,direct	累加器"或"到直接寻址;direct←(direct)∨(A)	2	1
ORL　direct,♯data	43,direct,data	立即数"或"到直接寻址;direct←(direct)∨data	3	2
XRL　A,♯data	64,data	立即数"异或"到累加器;A←(A)⊕data	2	1
XRL　A,direct	65,direct	直接寻址"异或"到累加器;A←(A)⊕(direct)	2	1
XRL　A,Rn	68～6F	立即数"异或"到累加器;A←(A)⊕(Rn)	1	1
XRL　A,@Ri	66～67	间接寻址 RAM"异或"累加器;A←(A)⊕((Ri))	1	1
XRL　direct,A	62,direct	累加器"异或"到直接寻址;direct←(direct)⊕(A)	2	1
XRL　direct,♯data	63,direct,data	立即数"异或"到直接寻址;direct←(direct)⊕data	3	2
CLR　A	E4	累加器清零;A←0	1	1
CPL　A	F4	累加器求反;A←($\overline{A}$)	1	1
RL　A	23	累加器循左移;$A_{n+1}←A_n,A_7←A_0$	1	1

(续表)

指令助记符	机器码	说明	字节数	机器周期数
RLC A	33	经过进位位的累加器循环左移；$A_{n+1} \leftarrow A_n, A_0 \leftarrow CY, CY \leftarrow A_7$	1	1
RR A	03	累加器右移；$A_n \leftarrow A_{n+1}, A_0 \leftarrow A_7$	1	1
RRC A	13	经过进位位的累加器循环右移；$A_{n+1} \leftarrow A_n, A_7 \leftarrow CY, CY \leftarrow A_0$	1	1

附 B 表 4 控制转移表指令

指令助记符	机器码	说明	字节数	机器周期数
LJMP addr16	02,addr16	长转移；PC←(PC)+3，PC←addr16	3	2
AJMP addr11	*	绝对转移；PC←(PC)+2，$PC_{10\sim 0} \leftarrow$ addr11	2	2
SJMP rel	80,rel	短转移(相对偏移)；PC←(PC)+2+rel	2	2
JMP @A+DPTR	73	相对 DPTR 的间接转移；PC←(A)+(DPTR)	1	2
JZ rel	60,rel	累加器为 0 则转移，为 1 则顺序执行下一条指令； 若(A)≠0，则 PC←(PC)+2； 若(A)=0，则 PC←(PC)+2+rel	2	2
JNZ rel	70,rel	累加器不为 0 则转移，为 0 则顺序执行下一条指令； 若(A)=0，则 PC←(PC)+2； 若(A)≠0，则 PC←(PC)+2+rel	2	2
CJNE A,#data,rel	B4,data,rel	比较立即数和 A 不相等则转移，相等则顺序执行下一条指令； 若(A)=(data)，则 PC←(PC)+3，CY←0； 若(A)>(data)，则 PC←(PC)+3+rel，CY←0； 若(A)<(data)，则 PC←(PC)+3+rel，CY←1	3	2
CJNE A,direct,rel	B5,direct,rel	比较直接寻址字节和 A 不相等则转移，相等则顺序执行下一条指令； 若(A)=(direct)，则 PC←(PC)+3，CY←0； 若(A)>(direct)，则 PC←(PC)+3+rel，CY←0； 若(A)<(direct)，则 PC←(PC)+3+rel，CY←1	3	2

（续表）

指令助记符	机器码	说 明	字节数	机器周期数
CJNE Rn,#data,rel	B8~BF,data,rel	比较立即数和寄存器不相等则转移,相等则顺序执行下一条指令; 若(Rn)=(data),则 PC←(PC)+3,CY←0; 若(Rn)>(data),则 PC←(PC)+3+rel,CY←0; 若(Rn)<(data),则 PC←(PC)+3+rel,CY←1	3	2
CJNE @Ri,#data,rel	B6~B7,data,rel	比较立即数和间接寻址 RAM 不相等则转移,相等则顺序执行下一条指令; 若((Ri))=(data),则 PC←(PC)+3,CY←0; 若((Ri))>(data),则 PC←(PC)+3+rel,CY←0; 若((Ri))<(data),则 PC←(PC)+3+rel,CY←1	3	2
DJNZ direct,rel	D5,direct,rel	直接寻址字节减 1 不为 0 则转移,为 0 则顺序执行下一条指令; direct←(direct)−1;若(direct)≠0,则 PC←(PC)+3; 若(direct)≠0,则 PC←(PC)+3+rel	3	2
DJNZ Rn,rel	D8~DF,rel	寄存器减 1 不为零则转移,为 0 则顺序执行下一条指令; Rn←(Rn)−1;若(Rn)≠0,则 PC←(PC)+2; 若(Rn)≠0,则 PC←(PC)+2+rel	2	2
LCALL addr16	12,addr16	长调用子程序; PC←(PC)+3,SP←(SP)+1; SP←(PC)$_L$,SP←(SP)+1; (SP)←(PC)$_H$,PC←addr16	3	2
ACALL addr11	*	绝对调用子程序; PC←(PC)+2,SP←(SP)+1; SP←(PC)$_L$,SP←(SP)+1; (SP)←(PC)$_H$,PC$_{10\sim 0}$←addr11	2	2
RET	22	从子程序返回; PC$_H$←((SP)),SP←(SP)−1; PC$_L$←((SP)),SP←(SP)−1;	1	2

(续表)

指令助记符	机器码	说　明	字节数	机器周期数
RETI	32	从中断返回； $PC_H \leftarrow ((SP)), SP \leftarrow (SP)-1$； $PC_L \leftarrow ((SP)), SP \leftarrow (SP)-1$	1	2
NOP	00	空操作	1	1

附 B 表 5　位操作指令

指令助记符	机器码	说　明	字节数	机器周期数
MOV C,bit	A2,bit	直接地址位送入进位位；$CY \leftarrow (bit)$	2	1
MOV bit,C	92,bit	进位位送入直接地址位；$bit \leftarrow CY$	2	2
CLR C	C3	清进位位；$CY \leftarrow 0$	1	1
CLR bit	C2,bit	清直接地址位；$bit \leftarrow 0$	2	1
CPL C	B3	进位位求反；$CY \leftarrow \overline{CY}$	1	1
CPL bit	B2,bit	直接地址位求反；$bit \leftarrow \overline{bit}$	2	1
SETB C	D3	置进位位；$Cy \leftarrow 1$	1	1
SETB bit	D2,bit	置直接地址位；$bit \leftarrow 1$	2	1
ANL C,bit	82,bit	进位位和直接地址位相"与"； $CY \leftarrow (CY) \wedge (bit)$	2	2
ANC C,$\overline{bit}$	B0,bit	进位位和直接地址位的反码相"或"； $CY \leftarrow (CY) \wedge (\overline{bit})$	2	2
ORL C,bit	72,bit	进位位和直接地址位相"与"； $CY \leftarrow (CY) \vee (bit)$	2	2
ORL C,$\overline{bit}$	A0,bit	进位位和直接地址位的反码相"或"； $CY \leftarrow (CY) \vee (\overline{bit})$	2	2
JC rel	40,rel	进位位为 1 则转移，为 0 则顺序执行下一条指令； 若$(CY)=0$，则 $PC \leftarrow (PC)+2$； 若$(CY)=1$，则 $PC \leftarrow (PC)+2+rel$	2	2
JNC rel	50,rel	进位位为 0 则转移，为 1 则顺序执行下一条指令； 若$(CY)=1$，则 $PC \leftarrow (PC)+2$； 若$(CY)=0$，则 $PC \leftarrow (PC)+2+rel$	2	2

（续表）

指令助记符	机器码	说　明	字节数	机器周期数
JB bit, rel	20, bit, rel	直接地址位为 1 则转移，为 0 则顺序执行下一条指令； 若(bit)=0,则 PC←(PC)+3; 若(bit)=1,则 PC←(PC)+3+rel	3	2
JNB bit, rel	30, bit, rel	直接地址位为 0 则转移，为 1 则顺序执行下一条指令； 若(bit)=1,则 PC←(PC)+3; 若(bit)=0,则 PC←(PC)+3+rel	3	2
JBC bit, rel	10, bit, rel	直接地址位为 1 则转移，并且该位清 0；直接地址位为 0 则顺序执行下一条指令； 若(bit)=0,则 PC←(PC)+3; 若(bit)=1,则 PC←(PC)+3+rel, bit←0	3	2

附 B 表 6　影响标志的指令

指　令	CY	标志 OV	Ac
ADD	√	√	√
ADDC	√	√	√
SUBB	√	√	√
MUL	0	√	
DIV	0	√	
DA	√		
RRC	√		
RLC	√		
SETB C	1		
CLR C	0		
CPL C	√		
ANL C, bit	√		
ANL C, /bit	√		
OR C, bit	√		
OR C, /bit	√		
MOV C, bit	√		
CJNE	√		

附录 C 单片机伪指令

在对由汇编语言编写的源程序进行汇编时,有一些控制汇编用的特殊指令,这些指令并不属于单片机指令系统,不产生机器代码,因此称为伪指令。伪指令只是在计算机将汇编语言转换为机器码时,用于指导汇编过程,告诉汇编程序如何汇编。下面介绍 MCS-51 系列单片机汇编程序常用的伪指令。

1. 起始地址伪指令 ORG

指令格式:ORG　nn(绝对地址或标号)

指令功能:ORG 伪指令总是出现在每段源程序或数据块的开始,它指明此语句后面的程序或数据块的起始地址。在汇编时由 nn 确定此语句后面第一条指令(或第一个数据)的地址。该段源程序(或数据块)就连续存放在之后的地址内,直到遇到另一个 ORG　nn 语句为止。

2. 汇编结束伪指令 END

指令格式:END

指令功能:END 伪指令是一个结束标志,用来指示汇编语言源程序段在此结束。因此,在一个源程序中只允许出现一个 END 语句,并且它必须放在整个程序(包括伪指令)的最后面,是源程序模块的最后一个语句。如果 END 语句出现在中间,则汇编程序将不汇编 END 后面的语句。

3. 字节数据定义伪指令 DB

指令格式:[标号:]　DB 字节常数或字符或表达式

指令功能:从指定的地址单元开始定义若干个字节的数值或 ASCII 码字符,各数据之间用逗号分隔,常用于定义一个数据表格存入程序存储器。在表示 ASCII 码字符时需要在字符上加单引号,标号表示数据表的首地址。

DB 定义的数据表一行可以写多个数据,当一行写不下要分行时,在下一行也必须用 DB 伪指令开头。

4. 字数据定义伪指令 DW

指令格式:标号:DW　字数据表

指令功能:从标号指定的地址单元开始,在程序存储器中定义字数据。该伪指令将字或字表中的数据根据从左往右的顺序依次存放在指定的存储单元中。

应特别注意:16 的二进制数,高 8 位存放在低地址单元,低 8 位存放在高地址单元。

5. 空间定义伪指令 DS

指令格式:标号:DS　表达式

指令功能：从标号指定的地址单元开始，在程序存储器中保留由表达式所指定的个数的存储单元作为备用空间，并都填以零值。

6. 伪指令 EQU

指令格式：符号名 EQU 表达式

指令功能：EQU 伪指令的功能是将表达式的值或特定的某个汇编符号定义为一个指定的符号名。使用 EQU 伪指令给一个符号名赋值后，这个标号在整个源程序中的值是固定的。也就是说在一个源程序中，任何一个符号名只能赋值一次。

7. 地址符号定义伪指令 BIT

指令格式：位名称　BIT　位地址表达式

指令功能：将位地址赋给指定的位名称。

参考文献

[1] 李庭贵．单片机应用技术及项目化训练[M]．成都:西南交通大学出版社,2009．
[2] 张迎辉．单片机实训教程[M]．北京:北京大学出版社,2005．
[3] 夏继强．单片机实验与实践教程[M]．北京:北京航空航天大学出版社,2001．
[4] 吴金戌,沈庆阳．8051单片机实践与应用[M]．北京:清华大学出版社,2002．
[5] 张宏伟,李德新．单片机应用技术实训[M]．北京:北京理工大学出版社,2010．
[6] 彭冬明,韦友春．单片机实验教程[M]．北京:北京理工大学出版社,2007．
[7] 袁秀英,李珍．单片机原理与应用教程[M]．北京:北京航空航天大学出版社,2006．
[8] 张旭涛,曾现峰．单片机原理与应用[M]．北京:北京理工大学出版社,2007．
[9] 徐玮,沈建良．单片机快速入门[M]．北京:北京航空航天大学出版社,2008．
[10] 李全利．单片机原理及应用技术[M]．北京:高等教育出版社,2004．
[11] 周坚．单片机轻松入门[M]．北京:北京航空航天大学出版社,2004．
[12] 李秀忠．单片机应用技术[M]．北京:人民邮电出版社,2007．
[13] 龚运新,胡长胜．单片机应用技术教程[M]．北京:北京师范大学出版社,2005．